Nutrition and Fitness for Athletes

World Review of Nutrition and Dietetics

Vol. 71

Series Editor

Artemis P. Simopoulos, Washington, D.C.

KARGER

Basel · Freiburg · Paris · London · New York · New Delhi · Bangkok · Singapore · Tokyo · Sydney

Nutrition and Fitness for Athletes

Volume Editors

Artemis P. Simopoulos
The Center for Genetics, Nutrition and Health,
American Association for World Health, Washington, D.C., USA

Konstantinos N. Pavlou
Hellenic Sports Research Institute, Olympic Athletic Center of Athens, Greece

19 figures and 33 tables, 1993

Basel · Freiburg · Paris · London · New York · New Delhi · Bangkok · Singapore · Tokyo · Sydney

World Review of Nutrition and Dietetics

Library of Congress Cataloging-in-Publication Data
International Conference on Nutrition and Fitness (2nd : 1992 : Athens, Greece)
Nutrition and fitness for athletes / 2nd International Conference on Nutrition and Fitness,
Athens, May 23–25, 1992 ; volume editors, Artemis P. Simopoulos, Konstantinos N. Pavlou.
(World review of nutrition and dietetics ; vol. 71)
Includes bibliographical references and index.
1. Athletes – – Nutrition.   2. Nutrition.   3. Physical fitness.
I. Simopoulos, Artemis P., 1933–   .   II. Pavlou, Konstantinos N.   III. Title.   IV. Series.
[DNLM: 1. Sports – – congresses.   2. Diet – – congresses.   3. Nutrition – – congresses.
4. Physical Fitness – – congresses.   W1 W0898 v.71 1993   QT 260 I538n 1992)
ISBN 3–8055–5707–8 (alk. paper)

Bibliographic Indices
This publication is listed in bibliographic services, including Current Contents® and Index
Medicus.

Drug Dosage
The authors and the publisher have exerted every effort to ensure that drug selection and dosage set forth in this text are in accord with current recommendations and practice at the time
of publication. However, in view of ongoing research, changes in government regulations, and
the constant flow of information relating to drug therapy and drug reactions, the reader is
urged to check the package insert for each drug for any change in indications and dosage and
for added warnings and precautions. This is particularly important when the recommended
agent is a new and/or infrequently employed drug.

Contents

Panel I

The Contribution of Macronutrients to Peak Performance of the Elite Athlete

Panel II

Concerns for Specific Population Groups in Relation to Nutrition and Fitness

Special Presentations

Appendix

Poster Abstracts

Panels III–VII are published in volume 72 of this series.

Conference Organization

Organized by

The Center for Genetics, Nutrition and Health
The Hellenic Sports Research Institute (HSRI)
The Olympic Athletic Center of Athens, Spyros Louis

Under the Patronage of

The General Secretariat of Athletics of Greece –
  Ministry of Culture/International Olympic Academy (IOA)
Food and Agriculture Organization of the United Nations (FAO)
World Health Organization (WHO)

Sponsors

General Secretariat of Athletics of Greece
The President's Council on Physical
  Fitness and Sports, USA
Pan American Health Organization
  (PAHO)
National Institute of Child Health and
  Human Development, National
  Institutes of Health (NICHD-NIH)
Hellenic Pharmaceutical Association

International Union of Nutritional
  Sciences (IUNS)
American Association for World Health
  (AAWH)
Amway Corporation
Kellogg Company
'Slim Fast' Nutritional Foods
  International, Inc.

Additional Support for the Conference and the Proceedings Provided by

National Institute of Alcohol Abuse and
  Alcoholism, NIH
Mead Johnson Nutritional Group/
  Bristol-Myers
The NutraSweet Company

Eastman Chemical Products, Inc.
Henkel Corporation
McNeil Specialty Products Company
Campbell Soup Company

Honorary Chairman

Kyriakos Virvidakis, MD (Greece)

Executive Committee

Artemis P. Simopoulos, MD (USA),
  Chairman
Konstantinos N. Pavlou, ScD (Greece),
  Co-Chairman
Per-Olof Astrand, MD (Sweden)
George L. Blackburn, MD, PhD (USA)
Peter Bourne, MD (USA)
Ji Di Chen, MD (China)
Nicholas T. Christakos (USA)
J.E. Dutra-de-Oliveira, MD (Brazil)
Nikos Filaretos, IOA (Greece)

Frank Kotsonis, PhD (USA)
Eleazar Lara-Pantin, MD (Venezuela)
Alexander Leaf, MD (USA)
John Lupien (FAO)
Wim H.M. Saris, MD, PhD
  (The Netherlands)
Leonidas Seitanides, MD (Greece)
A. Stewart Truswell, MD (Australia)
Kihumbu Thairu, PhD (UK, Kenya)
Clyde Williams, PhD (UK)

Conference Planning Committee

Artemis P. Simopoulos, MD (USA),
  Chairman
Konstantinos N. Pavlou, ScD (Greece),
  Co-Chairman
George L. Blackburn, MD, PhD (USA)
Ji Di Chen, MD (China)
Carlos Hernan Daza, MD, PhD (PAHO)
J.E. Dutra-de-Oliveira, MD (Brazil)
Nikos Filaretos, IOA (Greece)
Eleni Grigoriadou, MD (Greece)
Anthony Kafatos, MD (Greece)
Demetre Labadarios, MD, PhD
  (South Africa)

Eleazar Lara-Pantin, MD (Venezuela)
Alexander Leaf, MD (USA)
John Lupien (FAO)
Evangelia Maglara-Katsilambrou (Greece)
Paul Nestel, MD (Australia)
York E. Onnen (USA)
Jana Pařízková, MD, PhD, DSc
  (Czechoslovakia)
Victor Rogozkin, DBS (Russia)
Leonidas Seitanides, MD (Greece)
Stamatis Skoutas, MD (Greece)
Kihumbu Thairu, PhD (UK, Kenya)
Clyde Williams, PhD (UK)

Hellenic Organizing Committee

Leonidas Seitanides, MD, President
Eustathios Matsoukas, MD, Vice President

Konstantinos N. Pavlou, ScD, Secretary
George Rontoyannis, MD, Treasurer

Members

Dimitrios Chasiotis, PhD
Eleni Grigoriadou, MD
Evangelia Maglara-Katsilambrou, RD
Antonios Kafatos, MD
Panagiotis Kontoulakos, MD

Panagiota Markou, RD
Stamatios Skoutas, MD
Panagiotis Stamatopoulos, MD
Alexandros Tsopanakis, PhD

Dedication

The proceedings of the conference are dedicated
to the concept of positive health as enunciated by the
Hippocratic physicians (5th century BC).

*Positive health requires a knowledge of man's primary constitution
(which today we call genetics) and of the powers of various foods,
both those natural to them and those resulting from human skill
(today's processed food).*
But eating alone is not enough for health.
There must also be exercise,
of which the effects must likewise be known.
The combination of these two things makes regimen,
when proper attention is given to the season of the year,
the changes of the winds,
the age of the individual and the situation of his home.
If there is any deficiency in food or exercise the body will fall sick.

Preface

In May 1988, the First International Conference on Nutrition and Fitness was held at the International Olympic Academy (IOA) at Ancient Olympia under the patronage of the International Olympic Committee (IOC) and the World Health Organization (WHO). It was considered most appropriate to hold the conference at Ancient Olympia because the importance of nutrition and physical fitness to health were first conceived in Greece and implemented at Ancient Olympia. The International Olympic Academy works under the auspices of the IOC as the only recognized Olympic academy. The IOA operates under the supervision of the Hellenic Olympic Committee for the promotion of the Olympic education. It is the spiritual center of the Olympic movement. Olympism is a concept with roots going back to ancient Greece; it contains fundamental aims for the education of man, and according to a phrase of Baron Pierre Coubertin, founder of the modern Olympic Games in 1896, 'Olympism tries to concentrate in a luminous beam all the principles that contribute to the perfection of man'. Ancient Greek philosophers believed that physical beauty, strength and health were not the only attributes men and women should have; however, when combined with moral and spiritual virtues and promoted through exercise and contests, these attributes can create a perfect man, a well-balanced being in all his elements and actions. Thus, through the academy, Olympia is the permanent center of Olympism and plans to widen its research work in parallel to its educational activities and participation of scientists from all disciplines. Thus, holding the First International Conference on Nutrition and Fitness reestablished nutrition as an important aspect of Olympism as it was in ancient times.

At the completion of the First International Conference on Nutrition and Fitness the following resolutions were made [1]:

(1) The participants of the conference wish to encourage governments to develop programs related to better nutrition and improved fitness.

(2) Nutrition policies should be coordinated with programs to improve physical fitness.

(3) Programs should take into consideration the variations in need in relation to different age groups and social circumstances for guidance about dietary needs and physical activity.

(4) IOC and WHO should be leaders in stimulating and providing guidance.

(5) We should meet in Olympia every 4 years before the Olympic games to update advice in the light of research results. We should continue to stimulate governments to develop and maintain programs on nutrition and fitness.

The Second International Conference on Nutrition and Fitness was held on May 23–25, 1992 at the Olympic Athletic Center of Athens 'Spyros Louis' under the patronage of the IOA, the Food and Agriculture Organization of the United Nations (FAO) and the World Health Organization (WHO). Seven hundred and eighty scientists and policy makers from academia, industry and government from 29 countries representing the 5 continents of Europe, Australia, North and South America, Asia and Africa attended the conference. The conference consisted of seven panels: Panel I: The Contribution of Macronutrients to Peak Performance of the Elite Athlete; Panel II: Concerns for Specific Population Groups in Relation to Nutrition and Fitness; Panel III: Diet and Exercise in Cardiovacular Disease; Panel IV: Obesity; Panel V: Osteoporosis; Panel VI: New Concepts on Nutrition, Fitness and RDAs; Panel VII: Update on Policies and Programs in Nutrition and Physical Fitness; and a special presentation on 'Exercise and Mood'.

The honorary chairman of the conference was Dr. Kyriakos Virvidakis, General Secretary of Athletics of Greece, who officially opened the conference and said that he was 'looking forward to the conference recommendations and the Declaration of Olympia on Nutrition and Fitness'. Dr. Virvidakis expressed his hope 'to see everyone again at the Third International Conference on Nutrition and Fitness in Greece'. Dr. Leonidas Seitanides, President of the Olympic Athletic Center of Athens, welcomed everyone and affirmed his belief in the concepts of nutrition and fitness being essential for health. John Lupien, Director of the Food Policy and Nutrition Division of the FAO said, 'In preparing for the International Conference on Nutrition (ICN) to be held in December 1992, FAO and

WHO have become increasingly aware of the need for continued research into many different aspects of nutrition, dietary habits, exercise and lifestyles and the interrelationship between these factors. It is clear that the results of such research need wide dissemination and application and conferences of the type we are attending today are one of the most effective means to achieve this. In ICN preparation we have contacted many different organizations to obtain research information and different points of view to arrive at a consensus of the priority activities that need to be carried out in different countries. I look forward to hearing and participating in the presentations and deliberations on the many different aspects of nutrition and fitness which will take place in this conference over the next several days. I can assure you that the recommendations of this meeting will be a very timely addition to the ICN preparatory process and will be used in preparing the final ICN declaration and plan of action.'

Dr. Carlos Daza, representing WHO said, 'In this second International Conference, the World Health Organization is again pleased to grant its patronage to the International Conference on Nutrition and Fitness, jointly with the International Olympic Academy and the Food and Agriculture Organization of the United Nations. The first International Conference, held in Ancient Olympia in 1988, underlined the role of nutrition and fitness – as it was perceived by the Ancient Greeks – as part of the Olympic movement that has been for two millennia a vigorous stimulus for the physical and mental well-being of the human race. This Second International Conference on Nutrition and Fitness is highly relevant to the World Health Organization, for the member states are increasingly concerned about the ongoing nutrition and epidemiologic transitional trends, closely related to excessive or unbalanced dietary intake and sedentary habits. Both the objectives and topics of this conference are indeed relevant to WHO's priorities for healthy nutrition and lifestyle throughout the world, particularly with respect to the prevention of cardiovascular diseases, osteoporosis, obesity, cancer and functional consequences of erroneous diets and unhealthy behavior still affecting large groups of society. We look forward to the 'Declaration of Olympia on Nutrition and Fitness' as a corollary of this conference, based on the analysis of progress and accomplishments in nutrition and fitness made by your countries in the areas of research, training and program development. The World Health Organization is entirely willing to support renewed efforts and commitments of the international community and national governments to assure better nutrition for all people, reduction of hunger and malnutrition of vulnerable groups, elimination of micronutrient malnutrition and prevention of diet-related chronic and degenerative diseases'.

The conference cochairmen, Dr. Artemis P. Simopoulos and Dr. Konstantinos N. Pavlou thanked the conference participants and said that 'We will meet again in 1996, at which time we hope to have made considerable progress in having more countries participate in developing programs on nutrition and fitness'. This conference, as was the case with the first and all subsequent conferences, is dedicated to the concept of *positive health* as stated by the Hippocratic physicians.

Following the meeting in Athens, on May 26–27, 1992, a group of program participants from the Executive Committee and the Planning Committee representing Australia, Brazil, Greece, Kenya, Russia, South Africa, Sweden, UK, USA, Colombia and Venezuela, from the World Health Organization, the Pan American Health Organization, the Food and Agriculture Organization of the United Nations, the International Union of Nutritional Sciences, the Latin American Nutrition Society, the President's Council on Physical Fitness and Sports (USA), the Olympic Sports Center of Athens, The Center for Genetics, Nutrition and Health (USA), and other international organizations met at Ancient Olympia to develop the 'Declaration of Olympia on Nutrition and Fitness' [see pp. 1–8, this vol.].

As is customary, the group was welcomed by the representative of the IOA who officially opened the meeting with everyone listening to the Olympic hymn:

Immortal spirit of antiquity,
Father of the true, beautiful and good,
Descend, appear, shed over us thy light
Upon this ground and under this sky
Which had fits witnessed by unperishable fame.

Give life and animation to those noble games!
Throw wreaths of fadeless flowers to the victors
In the race and in the strife!
Create in our breasts, hearts of steel!

In thy light, plains, mountains and seas
Shine a roseate hue and form a vast temple
To which all nations throng to adore thee,
Oh immortal spirit of antiquity!

Official Olympic Hymn-Cantata by Costis Palamas
Set to Music by Spirou Samaras in 1896

The proceedings of the conference are published in two volumes in this series, volumes 71 and 72. Volume 71 is entitled 'Nutrition and Fitness for Athletes' and contains the papers from Panel I: The Contribution of Macronutrients to Peak Performance of the Elite Athlete; Panel II: Concerns for Specific Population Groups in Relation to Nutrition and Fitness; from Panel VI: New Concepts on Nutrition, Fitness and RDAs, the paper on 'Vitamin Requirements for Increased Physical Activity: Vitamin E'; from Panel VII: Update on Policies and Programs in Nutrition and Physical Fitness, the paper on 'Principles of Athletes' Nutrition in the Russian Federation'; the special presentation on 'Exercise and Mood'; and the conference poster abstracts. Volume 72 is entitled 'Nutrition and Fitness in Health and Disease' and contains the papers from Panel III: Diet and Exercise in Cardiovascular Disease; Panel IV: Obesity; Panel V: Osteoporosis; Panel VI: New Concepts on Nutrition, Fitness and RDAs; and Panel VII: Update on Policies and Programs in Nutrition and Physical Fitness.

The group felt strongly that the declaration of Olympia should be given wide distribution by having it published in major medical nutrition and sports journals in the languages of the participants. Everyone hoped that at the time of the Third International Conference on Nutrition and Fitness, many more countries will be participating and will have developed programs on Nutrition and Fitness.

These two volumes on nutrition and fitness should be of interest to practically everyone, but most definitely to physicians and scientists interested in the role of nutrition and fitness in health and performance, including nutritionists, exercise physiologists, geneticists, dietitians, educators, and policy makers (legislators and government officials).

Artemis P. Simopoulos, MD

Reference

1   Simopoulos AP (ed): Proceedings of the First International Conference on Nutrition and Fitness. Am J Clin Nutr 1989;49(suppl):909–1140.

Simopoulos AP, Pavlou KN (eds): Nutrition and Fitness for Athletes.
World Rev Nutr Diet. Basel, Karger, 1993, vol 71, pp 1–8

Declaration of Olympia on Nutrition and Fitness

May 26, 1992

The Second International Conference on Nutrition and Fitness met at the Olympic Athletic Center of Athens, May 23–25, 1992, under the patronage of the World Health Organization, the Food and Agriculture Organization of the United Nations, and the International Olympic Academy. Thirty-two scientific papers were presented to 780 registrants and 46 posters were displayed with results of recent investigations. A group of program participants met on May 26–27 at the International Olympic Academy, Ancient Olympia, to develop a declaration of aims and objectives resulting from the conference.

Present

Alexander Leaf (Cochairman), USA
Per-Olof Astrand (Cochairman), Sweden
Derek Prinsley (Secretary), Australia
Nicholas T. Christakos, USA
Carlos Hernan Daza, PAHO, WHO
Uri Goldbourt, Israel
Demetre Labadarios, South Africa
Eleazar Lara-Pantin, Venezuela
Meke Mukeshi, Kenya
J.E. Dutra-de-Oliveira, Brazil
York Onnen, USA
Konstantinos N. Pavlou, Greece
Eric Ravussin, USA (NIH)
Victor Rogozkin, Russia
Artemis P. Simopoulos, USA
Stewart Truswell, Australia
Clyde Williams, UK

Background

During the present century there have been unprecedented changes in life-style and patterns of health for humanity. Improved availability of wide varieties of food and less physical effort required for daily activities are prominent features of industrialized societies and of affluent groups in developing countries. However, humanity evolved during a more active, less-cushioned existence, with food often difficult to obtain. Present conditions in many affluent societies in which food choices are wide and physical activity is minimal may lead to poor nutrition with inadequate or incorrect food choices. Poor physical fitness results from a diminished need to be physically active. The interrelationship between nutrition and fitness is clear. The aims and objectives of the International Conference are to promote the advantages to health and fitness of good nutrition and regular physical activity for all stages of the human life span.

During more than 99% of our existence, human beings subsisted as hunters and food gatherers. During these 2 million years or more, our genetic code evolved to adapt us to a life-style as hunter-gatherers. Adaptations for survival were consonant with the need for habitual physical activity, including endurance and peak effort alternating with periods of rest and socialization. Major changes, however, have since taken place in both diet and physical activity. These changes began some 10,000 years ago with the development of agriculture and the domestication of animals. But increases in saturated fat from meat and dairy products, and trans fatty acids from margarines to present high levels, together with physical inactivity are much more recent phenomena, dating to the turn of this century. Even 10,000 years is too brief to allow significant adaptations to be made in our genetic code to these changes.

Human beings evolved on a diet that was not only low in saturated fats but was also balanced in the amounts of omega–6 and omega–3 fatty acids from seeds, green leafy vegetables, fish, and meat from animals and birds in the wild. The meat of wild animals and birds was lean, unlike the meat from domesticated, fattened cattle, sheep and fowl that human beings eat today. Thus, the diet was rich in vitamins and minerals and high in protein and fiber but low in fat. Vitamin C intake from wild green leafy vegetables and fruits was much higher than today's international recommendations and dietary allowances. The same was true for calcium and potassium, while the sodium content of the diet was lower than today's average intake.

Industrialized societies are characterized by an abundant and palatable food supply and by a sedentary existence at home, at work, and in transportation for most individuals. Today less than 1 % of energy used in farm and factory work comes from muscle power, whereas at the beginning of this century some 30 % of energy in these occupations came from human muscle power.

Children are naturally very active. They like to run and play, but today they are in schools (from nursery to kindergarten to primary and secondary schools) sitting most of the time. Few schools provide adequate programs or mandatory classes for physical exercise and individual involvement in sports. Instead, children have become spectators of sports rather than active participants in gymnastics, dance and informal play. Today the weight of a child may be predicted by the number of hours he or she watches television. Major increases in weight occur during adolescence and young adulthood with potentially dire consequences to health. Obesity is rampant in Western, developed societies and it is also occurring among the affluent in developing countries. Obesity is a major health problem of Western societies. It is apparently a health risk to change quickly from the life-style of a hunter-gatherer to that of a modern urban dweller. Insight into our biological heritage may help us to modify our current life-style in a positive way.

In most developing countries the nutritional problems are quite different. Dietary fat is already low and unrefined carbohydrate high but the intake of energy, protein, and micronutrients are all too often inadequate. A more bountiful and sanitary supply for all of the foods that are traditional in these cultures is needed. Surely, emulation of the excesses of the diets of Western, affluent societies is to be avoided.

We decry the existence of large numbers of hungry children and adults amidst the abundance of food in many of the industrialized countries. We call upon their governments to correct the maldistribution problems that allow such inhumane inequities to exist and to encourage food choices that provide optimal nutrition.

The Concept of Positive Health

In 480 BC, Hippocrates recognized the importance of the balance between food intake that provided fuel to the body (energy intake) and physical activity (energy expenditure) for health.

He developed the following concept of 'positive health':

Positive health requires a knowledge of man's primary constitution (which today we call genetics) and of the powers of various foods, both those natural to them and those resulting from human skill (today's processed food). But eating alone is not enough for health. There must also be exercise, of which the effects must likewise be known. The combination of these two things makes regimen, when proper attention is given to the season of the year, the changes of the winds, the age of the individual and the situation of his home. If there is any deficiency in food or exercise the body will fall sick.

Hippocrates also noted that 'death occurs earlier in the obese'. In addition, the Olympic ideal emphasized the need to be responsible for one's body, and the belief that a healthy mind resides in a healthy body. Greeks were proud of their athletes, but gymnastics were part of everyone's daily activities, they were not just for the few. Among the Greeks, the concept of positive health was important and occupied much of their thinking. Those who had the means and the leisure applied themselves to maintaining positive health, conceived aesthetically as an end in itself.

The Olympic Ideal

From ancient times, sport in its most competitive form – the Olympic Games – has been regarded and practiced as a kind of art and has affected art in a special way. Sport as a noble emulation, stimulating continuous self-improvement and striving for the best performance of the competing athletes, had inevitably an aesthetic aspect and inspired the plastic arts: painting, pottery, sculpture and architecture, as well as poetry, music and literature. The Olympic spirit did not concern only perfection of bodily movement, but also the whole of the human being as a psychosomatic unit. The fundamental principle of the Olympic ideal was excellence, i.e. the athlete had to excel over others and over his own performance. Physical education with its forms of play and Greek agonistics were part of the culture of the time.

The setting of our conference and its follow-up in Ancient Olympia prompted recollections of the Olympic ideal of ancient Greece and the Hippocratic concept of positive health. Ideally, the concept of positive health should be related to each individual's genetic characteristics and environment together with adequate food and exercise. This principle of positive health should apply to all people, not just be limited to athletes and the affluent few. The appeal of a nutritional and physical fitness pro-gram should be broad, with particular attention directed to children. If such a program is to succeed, there will have to be increased training of health professionals in nutrition, exercise physiology and genetics.

To promote an optimal life-style and successful aging, an important and urgent challenge is to teach and promote good nutrition and regular physical activity from childhood to old age, including advice about smoking, alcohol abuse and other recognized threats to health. The message needs to be clear and then it becomes a matter of personal choice for a better quality of life that can be abetted by regular exercise and better-informed eating habits.

Children and Adolescents

Children are our future and their role and effectiveness in society will be directly related to their individual and collective physical and mental health. All children must have the opportunity to start life with a potential for fitness based on optimal health. To attain this goal, the following are needed.

(a) All expectant mothers should have access to proper prenatal care and counseling on the importance of smoking cessation and avoidance of alcohol and addictive drugs to optimize the prognosis for their off-spring.

(b) All children should receive a healthful, nutritious diet without excesses or deficiencies of calories or essential nutrients. Instruction in nutrition and food should begin in childhood.

(c) All children should have the opportunity to engage in a variety of sports and physical fitness programs with emphasis on aerobic endurance, muscle strength, flexibility, and general fitness and to receive instruction in school about the health benefits of such exercise. Adoption of pleasurable forms of activity that can be enjoyed throughout life on a regular and frequent basis should be encouraged.

(d) All children and adolescents should be helped to avoid use of tobacco and abuse of alcohol and drugs and to abstain from early, unhealthful sexual practices.

(e) A positive self-image for all children should be fostered by universal access to education and opportunities to fulfill their potential in society.

(f) Universal access to health care should be provided for all children, with emphasis on primary care and health promotion to prevent illness and to achieve optimal health.

Promotion of good nutrition and regular aerobic physical activity for all throughout life must be a mainstay of any system that promotes the welfare of the individual and of society.

Physical Activity Improves Nutrition

Overweight due to excessive body fat is a major disorder of people in affluent and, increasingly, in some developing countries. Overweight increases the chance of developing chronic diseases including diabetes, high blood pressure, stroke, coronary heart disease, arthritis and possibly some forms of cancer. Ungainly appearance and reduced mobility hamper the individual's emotional well-being. Because obesity is often familial and has a strong genetic component, early identification of those at risk and proper counseling should aid in the prevention of obesity.

The two most effective measures to control overweight are regular physical activity and consumption of a low-fat diet. When people expend more energy through regular exercise and consume a wide variety of low-fat foods, the higher energy intake increases the probability that the resulting mixture of foods in their diet will meet requirements for all essential nutrients.

It should be emphasized that simple, nontaxing, physical activities, such as walking, cycling, swimming and gardening, are effective but must be regular, with a target of some such activity daily. Growing older in an environment where physical activity is encouraged and enjoyed can ensure continuation of such activities throughout life. Consequently, communities should be urged to provide opportunities and facilities particularly in cities, where open spaces and attractive worksite facilities will encourage physical activities for all age groups.

Performance

Good nutrition and regular physical exercise promote feelings of well-being and improve performance in all daily activities. For competing athletes energy intake must be adequate and consonant with levels of physical activity. Specific instruction is needed for adequate protein, high carbohydrate, low saturated fat, minerals, vitamin, and fluid intake. More research is needed on the nutritional requirements that will most effectively improve exercise capacity and performance.

Education, Development and Training

Because there is a dearth of training of health professionals in nutrition and physical education and a dereliction by them of responsibility for promoting this fundamental basis for good health, the public is bombarded by fads and conflicting pseudo-facts conceived by self-styled authorities. Information and education for the public is not provided in a reliable way. The opportunity for improved well-being through diet and physical activity is lost because the message has not been delivered effectively.

At present there is only minimal teaching in these disciplines of those who have educational and practical responsibilities for the health of the public. Therefore, there is a need to develop curricula in schools of nutrition, schools of physical education, and medical schools that will integrate nutrition, exercise physiology, and genetics into health education.

Declaration

(a) In developed countries technological developments have minimized physical activity, whereas variety and availability of foods make dietary choice a personal but not always well-advised decision.

(b) In most developing countries, the nutrition problems are quite different. Dietary fat is already low and unrefined carbohydrate high but the intake of energy, protein, and micronutrients are all too often inadequate. A more bountiful and sanitary supply of all the foods that are traditional in these cultures is needed. Surely, emulation of the excesses of the diets of Western affluent societies is to be avoided.

(c) The existence of large numbers of hungry children and adults amidst the abundance of food in many of the industrialized countries is destructive to the individual and to society. Governments must correct the maldistribution problems that allow such inhumane inequities to exist and encourage food choices that provide optimal nutrition for all.

(d) The adverse health effects of physical inactivity and consumption of high-fat diets have been repeatedly demonstrated in affluent societies by the high incidence of chronic diseases associated with these factors.

(e) Programs encouraging physical activity and good nutrition have now been shown to reduce diseases associated with inactivity and ill-advised diets, and can promote the quality of life.

(f) Understanding of the benefits to health from increased physical activity and good nutrition should be widely disseminated through extensive publicity.

(g) Health professionals should be educated in nutrition and exercise physiology to assume leadership roles in educating the public to the health benefits of physical activity and good nutrition.

(h) Education of the public should be promoted in schools at all levels, in the work place, through the media, and by health professionals.

(i) Advice provided to the public should be based on validated research findings in nutrition, genetics, and physiology. Research in these inter-related biomedical sciences deserves increased public and private support.

(j) Communities must provide clean and open spaces for children's playgrounds and adult sports and designate specific paths for pedestrians, cyclists, and other exercisers.

(k) Evidence is now convincing that general well-being and health can be greatly advanced by achievable adjustments of life-styles, nutrition, and physical activity. We call on all to respond.

The Contribution of Macronutrients to
Peak Performance of the Elite Athlete

Simopoulos AP, Pavlou KN (eds): Nutrition and Fitness for Athletes.
World Rev Nutr Diet. Basel, Karger, 1993, vol 71, pp 9–20

Energy Needs of the Elite Athlete

Konstantinos N. Pavlou

Hellenic Sports Research Institute, Olympic Athletic Center of Athens, Greece

Introduction

Energy needs and expenditure during athletic competition and/or training have been the subject of scientific investigations during the past decade, resulting in the accumulation of an extensive body of information in the international literature [1–14]. However, great controversy exists for the same sport activity and the different phases of training and competition.

Energy needs are supplied from the various foods and food supplements athletes consume. The contribution of the energy sources to the total energy pool is mainly determined by the nature, intensity and duration of the activity, as well as by the energy source availability. Additional factors are the gender of the athlete, his/her body weight, previous experience and training status, time of day, weather conditions and altitude. Overall, a qualitative and quantitative energy balance must be maintained on a day-to-day basis for athletes to compete optimally. A positive energy balance results in the accumulation of excessive body fat [15], while a negative energy balance leads to weight loss. In both cases optimal athletic performance is hindered.

When we attempt to classify the published studies on nutrition and sports, we notice that they can be separated into three distinct categories. In the *first category* we classify all studies that report energy intake. Although the authors use different titles such as 'nutrient intake', 'energy intake', 'dietary habits', 'dietary intake', 'nutritional assessment', etc. [1–10], all, in the final analysis, report the amount of kilocalories athletes consume. In the *second category* we can classify all studies reporting 'energy expenditure' [11, 12]. Here again, the authors use different terms, but the point is that they only deal with one aspect of nutrition, that of 'energy expenditure'. Finally, we notice a *third category* of studies where the authors report both 'energy intake', as well as 'energy expenditure'. In most of these studies the authors use the term 'energy balance' [13, 14].

In addition to energy 'intake', 'expenditure' and/or 'balance', most authors attempt to analyze and report the various micronutrients consumed, and compare them to the recommended dietary allowances (RDAs) [16]. However, none of these studies gives us the answer to the question of how many kilocalories an athlete needs, during competition and the various phases of training, for maintenance of an 'optimal body weight' and at the same time assist him/her to compete optimally.

What the Literature Can Tell Us

Using the information provided by the above-mentioned studies, it is very difficult, if not impossible, to assess the energy needs of the elite athlete. We must not undermine, however, the useful information the various published studies provide us, since it is important to know the dietary habits as well as the energy consumed and expended by athletes in various sport activities.

The information we get from these studies is not sufficient to help all those involved in counseling the coach and the athlete on how many kcal an athlete needs on a day to day basis. Those of us involved in sports nutrition know very well the problems we have to overcome. The importance of achieving 'optimal competitive body weight' and how energy consumed influences it cannot be emphasized enough. We should bear in mind that the difference of only 50 g of extra body weight might account for the variation in placement between the 1st and 6th place in competition, or not allow the athlete to compete in other cases at all.

The information available from the various studies is too generalized and does not represent the individual athlete at all. The existing information, if any, confuses the situation and misguides the athletic community.

Attempting to assess the information available in the literature, we compared the energy intake reported in various studies on male rowers and female gymnasts (table 1). We noticed a wide difference from study to study in mean energy consumed. Mean reported energy intake ranged from

Table 1. Reported energy intake in male rowers

De Wijn et al. [7]	$4,140 \pm 504$ kcal·day^{-1} or 46 kcal·kg^{-1}
Pavlou et al. [18]	$4,211 \pm 227$ kcal·day^{-1} or 48 kcal·kg^{-1}
Simonsen et al. [17]	$4,176 \pm 302$ kcal·day^{-1} or 53 hcal·kg^{-1}
Strauzenberg et al. [12]	$5,800$ kcal·day^{-1} or 57 kcal·kg^{-1}
Short and Short [4]	$5,267 \pm 315$ kcal·day^{-1} or 62 kcal·kg^{-1}
Ntimof [19]	$6,560$ kcal·day^{-1} or 75 kcal·kg^{-1}

Table 2. Reported energy intake in female gymnasts

Bernadot et al. [2]	elite athletes	$1,706 \pm 421$ kcal·day^{-1} or 56 kcal·kg^{-1}
Ntimof [19]		$2,580$ kcal·day^{-1} or 63 kcal·kg^{-1}
Grandjean [8]	elite athletes	$1,935 \pm 398$ kcal·day^{-1} or 70 kcal·kg^{-1}
Chen et al. [9]	amateur athletes	$1,637 \pm 199$ kcal·day^{-1} or 74 kcal·kg^{-1}
	elite athletes	$2,298 \pm 326$ kcal·day^{-1} or 51 kcal·kg^{-1}

46 kcal·kg^{-1}·day^{-1} [4, 7, 12, 17, 18] to 75 kcal·kg^{-1}·day^{-1} [19], a difference of 29 kcal·kg^{-1}·day^{-1}. Since mean body weight of the rowers was reported to be 88 kg, we are able to calculate the difference in energy intake reported to amount to 2,552 kcal·day^{-1} (88 kg $\times$ 29 kcal). The difference of 2,552 kcal·day^{-1} is too great to be taken lightly.

The situation is similar for female gymnasts (table 2). In this case mean reported energy intake ranged from 51 kcal·kg^{-1}·day^{-1} for the elite athletes, to 74 kcal·kg^{-1}·day^{-1} for the amateur gymnasts [2, 8, 9, 19]. The mean difference of 23 kcal·kg^{-1}·day^{-1} amounted to 713 kcal·day^{-1} (31 kg $\times$ 23 kcal), a great difference indeed for an athlete of such a small body weight.

It is obvious, therefore, that we cannot depend on the information obtained from these studies to assess energy needs for the athletes. The practice to assess daily energy intake through the widely used Dietary Recall is time consuming and has its problems [20]; and the information one gets does not represent the daily energy needs of the athlete, but rather, the energy consumed during the days of the investigation. At the same time the information obtained on 'energy expenditure' during training and/or competition cannot be generalized since the factors influencing energy expenditure in sports are many, and change from situation to situation.

Table 3. Equations most commonly used to estimate daily energy needs

		Men	Women
1	FAO equations [11]	15.4W − 27H + 717 = kcal	13.3W + 334H + 35 = kcal
2	Harris and Benedict [21]	13.7W − 5H + 6.8A = kcal	9.6W + 1.8H + 655 − 4.7A = kcal
3	Durnin and Passmore [23]	36.6W + 815 = kcal	31.1W + 580 = kcal

W = Body weight in kg; A = age in years; H = body height in meters for equation 1 and cm in equation 2.

Energy expenditure in nonathletic populations is determined by calculating resting energy expenditure (REE) from standard predictive equations [21], and then adding an additional 20% increment to account for daily routine physical activity. To this is added the diet-induced thermogenesis (DIT) which represents an average of 10% of the energy intake and its real value changes as a result of the diet composition. Alternatively, REE may be measured by indirect calorimetry [22]. In sports a third component is added, that of energy needs for training and/or competition.

Energy needs during training and/or competition have been thoroughly investigated and found to vary from athlete to athlete, and activity to activity. However, no data on REE exist, although its direct and/or indirect contribution to daily caloric needs may vary from 10 to 90% of the total [11, 13, 14]. Many equations are available in the literature [11, 21, 23]; however, none of them are suited for athletes since the populations from which they were derived did not include athletes and thereby do not represent them.

Methods

In an attempt to resolve the issue of accuracy in daily energy needs in athletes, REE was measured in male (n = 256) and female (n = 209) athletes by indirect calorimetry (mREE) and compared to the predicted REE (pREE) as calculated by the most widely used Harris and Benedict equation [21] (table 3).

Athletes reported to the laboratory in the postabsorptive state (8–12 h after the last meal) during the morning hours (08:00–10:00 a.m.). The subjects did not engage in any physical activity or smoke any cigarettes during that period. They rested for 30–45 min, lying in a dark, quiet and comfortable room. Thereafter, while the subjects rested supine in bed, mREE was measured every 20 s by indirect calorimetry, using the Horizon 2900 sensormedics metabolic chart [24]. The system measures oxygen consumption

Table 4. Anthropometric characteristics (mean ± SD)

	n	Age, years	Weight, kg	Height, cm	Fat, %
Males	256	21.5±4.5	76.8±14.6	180.3±10.5	12.0±5.4
Females	15	18.4±4.0	57.7±12	166.3±9.4	17.4±3.8

Table 5. Resting energy expenditure (kcal, mean ± SD)

	n	Predicted (HB)	Measured (IC)	$\overline{\Delta}$	p	% expected
Males	256	1,804±236	1,712±293	92±227	<0.001	95.2±12.3
Females	209	1,425±128	1,235±201	190±156	<0.001	86.4±11.2

HB = Harris-Benedict; IC = indirect calorimetry.

($VO_2 \cdot ml \cdot min^{-1}$) and carbon dioxide production ($CO_2 \cdot ml \cdot min^{-1}$). REE was automatically calculated by the abbreviated Weir formula [22] and expressed in $kcal \cdot day^{-1}$ by the following equation:

REE in $kcal \cdot day^{-1} = 3.94 \times VO_2 \, (l \cdot min^{-1}) + 1.1 \times VCO_2 \, (l \cdot min^{-1})$.

To assure the validity of the REE measurements, mean values achieved during the last 10 min of the steady state period were used in the calculations.

Comparisons between mREE and pREE were performed using a Student's two tailed t test for paired data, with a level of significance set at alpha = 0.05. Multiple regression analysis was used to describe the relationship between mREE and the predictive variables of height, age and weight in an effort to develop a useful predictor equation for application in athletes.

Mean age for the males was 21.5 ± 4.5 years and for the females 18.4 ± 4.0 years. The mean body weight was 76.8 ± 14.6 kg for the male athletes and 57.7 ± 12 kg for the females, while percent body fat was 12.0 ± 5.4 and 17.4 ± 3.8%, respectively (table 4).

Statistical analysis showed that the predicted values of REE were significantly greater (p < 0.001) than the measured values of resting energy expenditure in both male (1,804 ± 236 vs. 1,712 ± 293 kcal/day) and female athletes (1,425 ± 128 vs. 1,235 ± 201 kcal/day) (table 5). Measured REE (mREE) was found to be 95.2 ± 12 and 86.4 ± 11% of the expected for male and female athletes, respectively, while individual values ranged from 76 to 122% for male athletes (fig. 1) and from 74 to 126% for female athletes (fig. 2) of the expected normal values, with only 73% of the male athletes and 78% of the female

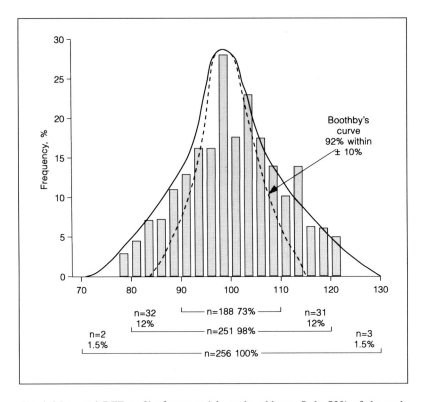

Fig. 1. Measured REE as % of expected in male athletes. Only 73% of the male athletes had measured resting energy expenditure that was within ± 10% of the expected as compared to 92% of Boothby's nonathletic population.

athletes having REE within ±10% of the expected as compared to 92% of Boothby's normal nonathletic population [25].

Multiple regression analysis with height, weight, and age as independent variables showed a statistically significant (p < 0.001) correlation for male (r = +0.74) and female (r = +0.63) athletes, respectively.

The regression equations resulting from the analysis were:

For male athletes:

REE kcal·day^{-1} = 12(W) + 6(H) − 8.5(A) − 106.

For female athletes:

REE kcal·day^{-1} = 11.1(W) + 0.7(H) − 7.7(A) + 624.

Where W = body weight in kg, H = body height in cm, and A = age in years.

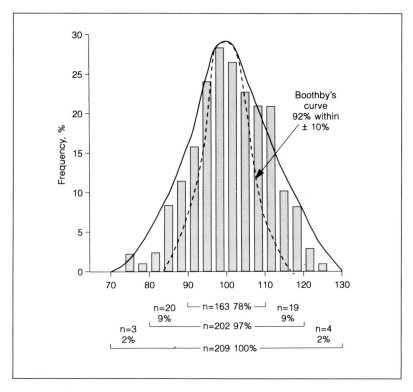

Fig. 2. Measured REE as % of expected in female athletes. Only 78% of the female athletes had measured resting energy expenditure that was within ± 10% of expected values, as compared to 92% of Boothby's nonathletic population.

Discussion

No explanation can be given for the observed reduced REE of our elite athletes when compared to the average, nonathletic, healthy population of the Harris-Benedict equation (HBE) study [21].

Although no data on body composition of the 136 males and 103 females studied by Harris and Benedict exist, we can safely assume that their subjects had greater total body fat than the population of the elite athletes who participated in this study, which had a mean body fat of only 12.0 ± 5.4% for the male, and 17.4 ± 3.8% for the female athletes (table 4).

Since it is well established that lean body mass (LBM) correlates highly with REE [26, 27], the population studied in this investigation ought to have greater REE than the HBE predicts, and not lower, as is evidenced by the findings of this study. Various reports in the literature indicate that REE per kg of LBM in nonathletic populations ranges from 28 to 32 kcal [26, 28], as compared to the reduced values in this study found to be 25.3 and 25.9 kcal/kg of LBM for the male and female athletes, respectively. It is well established that athletes, especially those involved in endurance activities, exhibit an increased efficiency when compared to nonathletic populations. Recent reports indicate that this is an adaptive response in novice athletes as well, when they initiate a regular training regimen [29, 30]. A similar phenomenon has been observed in obese male subjects undertaking a weight loss program by dieting and exercise [31].

The HBE has been found not to reflect the REE of morbidly [32] and moderately [28] obese subjects. A similar phenomenon has been found to hold for male and female swimmers as well [33]. Furthermore, the findings of this study indicate that the HBE is not suitable to be used for highly trained athletes as well.

Maintenance of ideal body weight requires a continuous daily energy balance. Although it seems to be fairly easy for the average individual to accomplish this, it represents a very difficult task for the athlete if one considers the fact that daily energy expenditure varies, since training intensity and duration changes from day to day, and season to season.

The athlete and coach must be provided with the necessary information which will help them calculate daily energy needs on a weekly basis. The different equations available in the literature are not suited for athletes, since no athletes were included in the population from which they were derived. From the many proposed equations, let us examine the three most commonly used and demonstrate the extent of the possibility of miscalculating daily energy needs for athletes (table 3).

In the first example (table 6) we will use data obtained from the Greek gold metal winner in the 82 kg category of weight lifting at the summer Olympic Games in Barcelona, Spain. Looking at the data we see that his daily energy needs (without energy needs for training and/or competition) as determined by indirect calorimetry were 2,188 kcal·day^{-1}. Comparing it to the values we get from the equations in our example, we can see that the Durnin and Passmore equation overestimates his daily needs by 1,738 kcal·day^{-1}, the HBE by 159 kcal·day^{-1} and the FAO equation for his age, by 187 kcal·day^{-1}.

Table 6. Daily energy needs for a male weightlifter[1] (body weight = 85 kg, height = 173 cm, age = 21 years, body fat = 7%)

	REE	DEN	$\overline{\Delta}$ from indirect calorimetry
FAO equation, kcal	1,979	2,375	187
Harris and Benedict equation, kcal	1,956	2,347	159
Durnin and Passmore equation, kcal	3,272	3,926	1,738
Indirect calorimetry, kcal	1,823	2,188	–
Proposed equation, kcal	1,774	2,129	–59

DEN = Daily energy needs without training and/or competition; REE = resting energy expenditure.
[1] 1992 Olympic Gold Medal winner.

One might speculate that the extra 159 kcal·day^{-1} the HBE overestimates is not the end of the world. However, we must not forget that a mere 159 kcal·day^{-1} accounts for more than 58,000 kcal per year, or 7.2 kg of extra body fat. To make the case stronger, it is worth mentioning that the athlete of our example, 2 months prior to the Olympic Games, competing in the European Championship against the same opponents and lifting the same weight, finished in fourth place simply because he was found to weigh 200, 100 and 50 g heavier than the first, second and third finisher, respectively. Looking at table 7 we see that the equation we propose underestimates his actual daily needs by a mere 59 kcal·day^{-1}, a difference easier to handle.

A similar situation is seen in the second example of a female sprinter (table 7). The athlete is the Greek gold metal winner of the 100 m hurdles race of the 1992 summer Olympic Games in Barcelona. The best predictive equation in this case is the FAO equation, which overestimates her energy needs by 135 kcal·day^{-1} (50,000 kcal/year) or 6.2 kg of extra fat per year. The new proposed equation underestimates her energy need by only 25 kcal·day^{-1}.

As illustrated in figure 1, 73% of the male athletes had resting metabolic rates within ± 10 of their expected values, whereas 92% of the healthy volunteers studied by Boothby [25] had resting metabolic rates within ± 10 of the predicted. The mREE of the weight lifter in our example (table 6) occurred at 101% of the expected. The proposed new equation

Table 7. Daily energy needs for a female sprinter[1] (body weight = 63 kg, height = 160 cm, age = 23, body fat = 11%)

	REE	DEN	$\overline{\Delta}$ from indirect calorimetry
FAO equation, kcal	1,424	1,709	135
Harris and Benedict equation, kcal	1,449	1,739	165
Durnin and Passmore equation, kcal	2,116	2,539	965
Indirect calorimetry, kcal	1,312	1,574	–
Proposed equation, kcal	1,291	1,549	–25

REE = Resting energy expenditure; DEN = daily energy needs without training and/or competition.
[1] 1992 Olympic Gold Medal winner in 100 m hurdles.

underestimates the athletes' daily needs (without energy needs for training and/or competition) by 2.8% (59 kcal·day^{-1}).

A similar phenomenon is shown in figure 2, where 78% of our female athletes had resting metabolic rates within ± 10 of the expected values, as compared to 92% of Boothby's volunteers. The daily energy needs (1,574 kcal) of the female athletes in our example (table 7) are underestimated by the new proposed equation by only 25 kcal·day^{-1} (1.6%).

Conclusions

The various equations available for calculating daily energy needs for athletes are not accurate enough, since no athletes were included in the populations from which they were derived. The ability to accurately predict daily energy needs for the elite athlete is very important, since training and proper nutrition represent the two most important determinants for successful competition.

The equations developed from this study accurately predict daily energy needs for both male and female elite athletes. Possible applications and ways to test their usefulness should be considered.

References

1   Hickson EJ, Risser LW, Johnson WC, Palmer R, Stockton AM, Duke EJ: Nutritional intake from food sources of high school football athletes. J Am Diet Assoc 1987;87:1656–1659.
2   Bernadot D, Schwarz M, Heller WD: Nutrient intake in young, highly competitive gymnasts. J Am Diet Assoc 1989;89:401–403.
3   Manone MM, Besenfelder DP, Wells C, Carroll SS, Hooker PS: Nutrient intake and iron status in female long-distance runners during training. J Am Diet Assoc 1989; 89:257–259.
4   Short HS, Short RW: Four-year study of University athletes dietary intake. J Am Diet Assoc 1983;82:632–645.
5   Parizkova J, Novak J: Dietary intake and metabolic parameters in adult men during extreme work load. World Rev Nutr Diet 1991;65:72–98.
6   Perron M, Endress J: Knowledge, attitudes and dietary practices of female athletes. J Am Diet Assoc 1985;85:573–576.
7   De Wijn FJ, Lensink J, Post BG: Diet, body composition and physical condition of champion rowers during periods of training and out of training. Bibl Nutr Dieta, Basel, Karger, 1979, vol 27, pp 143–148.
8   Grandjean CA: Macronutrient intake of US athletes compared with the general population and recommendations made for athletes. Am J Clin Nutr 1989;49:1060–1066.
9   Chen JD, Wang FJ, Zhao WY, Wang WS, Jiao Y, Hou YX: Nutritional problems and measures in elite and amateur athletes. Am J Clin Nutr 1989;49(suppl):1084–1089.
10   Heinemann L, Zerbes H: Physical activity, fitness, and diet: Behavior in the population compared with elite athletes in the GDR. Am J Clin Nutr 1989;49:1007–1016.
11   Parizkova J: Nutrition, energy expenditure and exercise. Proc 3rd Int Course Physiol Biochem Exerc Training, 1987, pp 239–255.
12   Strauzenberg ES, Schneider F, Donath R, Zerbes H, Kohler E: The problem of dieting in training and athletic performance. Bibl Nutr Dieta. Basel, Karger, 1979, vol 27, pp 133–142.
13   Hagerman FC, Hagerman MT, Falkel JE, Korzeniowski K, Oleary E: A comparison of energy input and output among elite rowers.
14   Rontoyannis GP, Skoulis T, Pavlou KN: Energy balance in ultramarathon running. Am J Clin Nutr 1989;49:976–979.
15   Bray AG: Obesity in America. NIH Publ No 80–359, 1980:37–68.
16   Food and Nutrition Board: National Research Council. Recommended Dietary Allowances, ed 9. Washington, National Academy of Sciences, 1980.
17   Simonsen CJ, Sherman MW, Lamb RD, Dernbach RA, Doyle AJ, Strauss R: Dietary carbohydrate, muscle glycogen, and power output during rowing training. J Appl Physiol 1991;70:1500–1505.
18   Papadokonstantaki M, Pavlou KN, Hassapidu M: Nutrient Intake of Greek Elite Athletes. World Rev Nutr Diet. Basel, Karger, 1993, vol 71, pp 183–184.
19   Ntimof E: Sports Nutrition. Jusantor, Sofia, 1987.

20   Guthrie AH, Crocetti FA: Variability of nutrient intake over a 3-day period. J Am Diet Assoc 1985;85:325–327.

21   Harris JA, Benedict FG: Biometric Studies of Basal Metabolism in Man. Washington, Carnegie Institute of Washington, 1919, Publ No 297.

22   Der Weir JB: New methods of calculating metabolic rate with special reference in protein metabolism. J Physiol 1949;109:1–9.

23   Durnin JV, Passmore R: Energy Work and Leisure. London, Heinemann Educational Books, 1967.

24   Metabolic Measurement Chart 2900: Sensormedics Corporation, 22705 Savi Ranch Parkway, Yorda, Calif., USA, 1991.

25   Boothby WM: In Dubois EF (ed): Basal Metabolism in Health and Disease, ed 3. Philadelphia, Lea & Febiger, 1936, pp 163–164.

26   Webb P: Energy expenditure and fat-free mass in men and women. Am J Clin Nutr 1981;34:1816–1826.

27   Halliday D, Hesp R, Stalley SF: Resting metabolic rate, weight, surface area and body composition in obese women. Int J Obes 1979;3:1–6.

28   Pavlou KN, Hoefer AM, Blackburn GL: Resting energy expenditure in moderate obesity. Ann Surg 1986;203:136–141.

29   Janssen GME, de Graef CJJ, Saris WHM: Food intake and body composition in novice athletes during a training period to run a marathon. Int J Sports Med 1989; 10:517–521.

30   Westerterp KR, Saris WHM: Limits of energy turnover in relation to physical performance achievement of energy balance on a daily basis. J Sports Sci 1991;9:1–15.

31   Pavlou KN: Metabolic adaptation in male obese subjects engaged in weight loss program by diet and exercise. Unpubl data.

32   Feurer ID, Crosby LO, Buzby GP, Rosato EF, Mullen JL: Resting energy expenditure in morbid obesity. Ann Surg 1983;197:17–21.

33   Pavlou KN: Resting energy profile of the elite Greek swimmer; in Parizkova J (ed): Nutrition, Metabolism and Physical Exercise. Prague, Charles University, 1989, pp 51–58.

Konstantinos N. Pavlou, DSc, Hellenic Sports Research Institute, Olympic
Athletic Center of Athens, 37 Kifissias Avenue, Maroussi 15123, Athens (Greece)

Simopoulos AP, Pavlou KN (eds): Nutrition and Fitness for Athletes.
World Rev Nutr Diet. Basel, Karger, 1993, vol 71, pp 21–33

Exercise and Protein Metabolism

William J. Evans

Human Physiology Laboratory, USDA Human Nutrition Research Center on
Aging, Tufts University, Boston, Mass., USA

Introduction

The substantial energy requirements of endurance exercise are primarily met by the oxidation of skeletal muscle glycogen and triglycerides as well as hepatic glycogen and adipocyte triglyeride stores as blood-borne glucose and free fatty acids. These fuels can account for 90–95% of the total energy requirement. However, the additional requirement is met by the oxidation of protein, which unlike fats and carbohydrate, is a non-renewable source of energy. This review will focus on the relationship between endurance and strengthening exercise and protein metabolism and the evidence for increased dietary protein requirements due to exercise. For the most part, two experimental approaches have been used. Early studies used the appearance of nitrogen-containing waste products, especially urea, in the blood and urine as an index of protein oxidation. Changes in the oxidation of individual amino acids using isotopes labeled as tracers have been employed more recently.

Submaximal Exercise and Protein Metabolism

The theory of the great German chemist, Justus Liebig, that protein was the primary fuel of working muscle [1], was tested by a number of 19th century scientists. Fick and Wisclicenus [2] hiked up the Faulhorn in 1865 and collected their urine for nitrogen analysis before and during the hike. Unfortunately, they confounded their results by placing themselves on a protein-free diet the day before their climb, thereby certainly reducing

Table 1. Nineteenth century studies on the effect of work on protein metabolism [from ref. 3]

Beigl (1855)	Increased N output following work on a meager diet, but even larger with protein-rich diet
Smith (1862)	Definite rise in N output on day of 29-mile treadmill walk and day following
Voit and Pettenkofer (1866)	No effect of work on N output
Fick and Wisclicenus (1866)	Decreased N output in work urine, post-work urine (but protein-free diet consumed and pre-work control urine reflects normal dietary protein level)
Parkes (1866–1871)	Slightly increased N output both on protein-free diet and mixed diet
Weigelin (1868)	Increased N output after 2 h hard work; rise most pronounced in post-work period
Schenk (1874)	Data shows slight but definite rise in urea output during work, initial post-work days, though he denied an effect
Pavy (1876)	Professional pedestrians on ample diet-increased N output on work days compared to rest days
Flint (1877)	Increased N output with 63.5 miles/day for 5 days
Breitzcke (1877)	Increased N output in convicts on work days
Argutinsky (1890)	Increased N output with work, present even when additional energy added bringing diet to adequate total energy supply
Paton (1891)	Small but definite rise on first and second post-days in a student consuming 3,979 kcal, 16 g N
Krummacher (1896)	Increased N output on post-work days at energy levels 38 and 64 kcal/kg, less when energy increased to 72 kcal/kg

their urinary nitrogen excretion. They also stopped their urine collection upon completion of the climb and, therefore, eliminated the detection of any post-exercise rise in urinary nitrogen loss. They concluded that protein oxidation provided only a small portion of the energy required to make their climb. As reviewed by Cathcart [3], and summarized in table 1, a slight increase in nitrogen excretion during and following work confirms the notion that protein oxidation does not provide the major source of energy for the working muscles.

The availability of carbohydrate as a fuel during exercise influences the oxidation of protein. Lemon and Mullen [4] estimated that the oxidation of protein (from serum, urine, and sweat urea losses) during prolonged submaximal exercise was double (up to 12% of the total energy demand) the glycogen-depleted state when compared to similar exercise in the glycogen-loaded state. Paradoxically, exercise in the heat (30 °C) is associated with a lower rate of protein utilization when compared to that seen during and for 2 days following similar exercise at 5 and 20 °C. Exercise in a hot environment has been demonstrated to increase the rate of glycogen utilization [5], a condition which should be associated with increased rates of protein utilization. Gender differences in substrate utilization during prolonged submaximal exercise have been seen. Tarnopolsky [6] and co-workers have shown that a group of female athletes matched to males by age and $\dot{V}O_{2max}$ used significantly less glycogen and protein (estimated from 24-hour urinary nitrogen excretion) than did men while running on a treadmill at 65% $\dot{V}O_{2max}$.

A number of studies have used the primed, constant infusion of ^{13}C-leucine to measure the rate of oxidation of this essential amino acid during exercise. These studies indicate that submaximal exercise does not alter leucine flux but substantially increases the rate of whole body leucine oxidation [7–9]. Rennie et al. [10] demonstrated that the increase in leucine oxidation rates was directly related to the intensity of the exercise. Knapick et al. [7] found that a complete 3.5-day fast did not cause an increase in leucine flux during exercise but caused a 44% increase in the rate of leucine oxidation. We [11] demonstrated that high-intensity exhaustive exercise caused a significant 48% accumulation of muscle α-ketoisocaproic acid (KIC). The elevation in muscle KIC concentration was not reflected in simultaneous changes in plasma KIC levels, suggesting a limited diffusion rate from muscle to blood. This study indicates that high-intensity, brief exercise is associated with accelerated transamination of leucine.

Devlin et al. [12] examined recovery from 3 h of cycling at 75% of $\dot{V}O_{2max}$ and found that whole body protein breakdown was not increased above resting levels, leucine oxidation was decreased, and nonoxidative leucine disposal (synthesis) was increased when compared to pre-exercise resting values. Using a much lower exercise intensity (4 h at 40% $\dot{V}O_{2max}$), Carraro et al. [13] also examined postexercise recovery. They found a significant increase in the muscle fractional synthetic rate during the recovery period.

The concentration of urea in plasma and urine increases during submaximal exercise and remains high for some time later, also in proportion to the intensity and duration of the exercise [14, 15]. The increased oxidation of indispensable amino acids during submaximal exercise must, therefore, increase the need for dietary protein.

Dietary Protein Requirements

While the experiments examining protein metabolism using ^{13}C-leucine (or other labeled amino acids) indicate increased oxidation during exercise, these studies do not directly show an increased requirement for dietary protein. Gontzea et al. [16] conducted a carefully controlled study in which 30 healthy young men consumed a diet containing 1.0 g protein/kg body weight. Nitrogen balance determinations were performed for three periods, a sedentary adaptation period, a 4-day exercise period, and a 4-day sedentary postexercise period. The daily exercise consisted of six 20-min intervals on a cycle ergometer at an intensity of 8–10 kcal/min, separated by 30-min breaks. Energy intake was adjusted during the exercise period to provide an extra 50 kcal/kg body weight$\cdot$day^{-1}. Sweat nitrogen losses were included in the calculation of nitrogen balance. The mean became negative during the exercise period and did not become positive even when the dietary protein intake was increased to 1.5 g$\cdot$kg$^{-1}\cdot$day^{-1}. A follow-up study examined the effect of a longer training period on nitrogen balance using a similar exercise load and a diet containing 1.0 g protein$\cdot$kg$^{-1}\cdot$day^{-1}. Nitrogen balance became negative with the onset of the exercise period but approached equilibrium by 2 weeks of training. The subjects in these studies were initially sedentary, and, therefore, the studies of Gontzea et al. [16, 17] do not address the question of whether athletes who have adapted to a high level of training intensity and duration have a high protein requirement when energy demands are met.

Tarnopolsky et al. [18] attempted to determine the protein requirements in bodybuilders and endurance trained men. These investigators examined nitrogen balance in these athletes on two different dietary protein intakes, both of which were above the amount required for achievement of balance. By extrapolating a line connecting the balance figures on two levels of protein intake, they estimated that bodybuilders required 1.12 times and endurance athletes required 1.67 times more daily protein than did sedentary controls. However, these protein requirements were extrapolated from nitrogen balance obtained with protein intakes of 1.7 and 2.65 g·kg^{-1}·day^{-1} and almost certainly overestimated protein requirements when compared with results that were closer to zero and included negative values.

Using nitrogen balance to estimate dietary protein requirement at three different dietary intakes (0.6, 0.9, and 1.2 g·kg^{-1}·day^{-1} of high quality protein over three separate 10-day periods), we [19] found that habitual endurance exercise was associated with dietary protein needs greater than the current Recommended Dietary Allowance of 0.8 g·kg^{-1}·day^{-1} and averaged 0.94 ± 0.05 g·kg^{-1}·day^{-1}. Whole-body protein turnover, using [^{15}N]glycine as a tracer, and 3-methylhistidine excretion were not different from values reported for sedentary men. Protein requirements expressed as a percent of energy needs (averaging 3,910 ± 240 kcal/day) showed that these subjects needed only 6.9 ± 0.5% of total dietary calories as protein. This suggests that well-trained individuals consuming an average American diet (12–15% protein) and adequate amounts of energy are not likely to have an inadequate dietary protein intake. However, we [20] examined the dietary intake of a group of eumenorrheic and amenorrheic athletes with similar exercise habits. There were no differences between the two groups in $\dot{V}O_{2max}$, number of miles run per week, or in body composition. When compared to the eumenorrheic athletes, the amenorrheic women reported consuming lower amount of calories (1,730 ± 152 vs. 2,250 ± 141 kcal) and protein (0.7 ± 0.1 vs. 1.0 ± 0.1 g·kg^{-1}·day^{-1}) implicating inadequate dietary calories and/or protein as a potential cause of athletic amenorrhea. When compared to the eumenorrheic women, the amenorrheic athletes have lower estradiol, estrone, LH (and LH pulse amplitudes), FSH, and T$_3$ levels [21], a hormone profile often seen in women on severely hypocaloric diets or women suffering from protein/calorie malnutrition.

Exercise-Induced Muscle Damage and Protein Metabolism

Muscle contraction and shortening produces a concentric action, however, when skeletal muscle lengthens as it produces force the result is an eccentric muscle action. An example of this is lifting a weight (concentric action) and lowering it (eccentric action). At the same power output, the oxygen cost of eccentric exercise is lower than that of concentric exercise [22]. Despite the lower oxygen cost, eccentric exercise has been demonstrated to be a potent cause of muscle damage [23, 24], DOMS, and increased circulating creatine kinase (CK) activity [25].

Running a marathon can cause extensive skeletal muscle damage [26, 27]. Warhol et al. [27] showed a characteristic pattern of muscle damage, with tearing of sarcomeres at the Z-band level followed by movement of fluid into the muscle cells in biopsies taken in the days following the race. Mitochondrial and myofibrillar damage showed progressive repair by 3–4 weeks after the marathon. Late biopsies (8–12 weeks after the race) showed central nuclei and satellite cells characteristic of a regenerative response. The damage seen by these investigators is very similar to the ultrastructural changes in skeletal muscle resulting from eccentric exercise.

The extent of the ultrastructural evidence of damage is greater well after the initial damaging exercise. Friden et al. [28] found more damaged muscle fibers 3 days after when compared to that seen only 1 h after high tension eccentric exercise. Newham et al. [23] also showed that eccentric exercise caused immediate damage, but that biopsies taken 24–48 h after the exercise had more marked damage. These data are indicative of an ongoing process of skeletal muscle repair consisting of increased degradation of damaged proteins and increased rate of protein synthesis.

Following only one bout of high-intensity eccentric exercise [25], previously sedentary men showed a prolonged increase in the rate of muscle protein breakdown, evidenced by an increase in urinary 3-methylhistidine/creatinine which peaked 10 days later. In addition, an increase in circulating interleukin-1 levels in these subjects was seen 3 h after the exercise. Endurance-trained men, performing the same exercise, did not display increased circulating IL-1 levels. However, their pre-exercise plasma IL-1 levels were significantly higher than those seen in the untrained subjects.

Damage to tissue as well as infection stimulates a wide range of defense reactions, known as the acute-phase response [29]. The acute-phase response is critical for its antiviral and antibacterial actions as well as promoting the clearance of damaged tissue and subsequent repair. Within hours of injury or exercise [30], the number of circulating neutrophils can increase many fold. Neutrophils migrate to the site of injury where they phagocytize tissue debris and release factors known to increase protein breakdown such as lysozymes and oxygen radicals [31]. Greater neutrophil increases have been observed after eccentric exercise than after concentric exercise [32]. While neutrophils have a relatively short half-life (1 or 2 days within tissue, [3]), the life span of monocytes may be 1–2 months after migration to damaged tissue [34]. Substantial monocyte accumulation in skeletal muscle was found after completion of a marathon. Following eccentric exercise, monocyte accumulation in muscle was not seen until 4–7 days later [35, 36]. In addition to the capability to phagocytize damaged tissue, monocytes secrete cytokines such as IL-1 and tumor necrosis factor (TNF). These and other cytokines mediate a wide range of metabolic events having an effect on virtually every organ system in the body.

Elevated cytokine levels during infection or injury have different and selective effects. IL-1 mediates an elevated core temperature during infection [37]. In laboratory animals, IL-1 and TNF increase muscle proteolysis and liberation of amino acids [38], possibly providing substrate for increased hepatic protein synthesis. While circulating IL-1 has been shown to increase acutely as a result of eccentric exercise [39], by 24 h after the exercise, it returned to resting levels. Biopsies of the vastus lateralis taken before, immediately after, and 5 days after downhill running, showed an immediate and prolonged increase in IL-1b [40]. This study implicates muscle IL-1b in the postexercise change in protein metabolism.

Eccentric exercise induced increases in muscle hydrolase activity [41], intracellular $[Ca^{2+}]$ [42], and IL-1b; and urinary 3-methylhistidine levels indicate that muscle protein turnover is also increased. Fielding et al. [43] used a primed, constant infusion of 1-^{13}C-leucine before, immediately after, and 10 days after a single bout of high-intensity eccentric exercise in previously sedentary old and young men to estimate whole body protein metabolism. Unlike the studies of Devlin et al. [12] and Carraro et al. [13] which used concentric exercise, Fielding et al. [43] found that leucine oxidation and flux were significantly elevated (compared to pre-exercise samples) at both postexercise time points, indicating a prolonged increase in protein turnover. These results tend to support those showing that previously sedentary young men consuming 1 g protein$\cdot$kg$^{-1}\cdot$day^{-1} showed an increased urinary nitrogen excretion and a prolonged period of negative nitrogen balance when beginning a vigorous exercise program [16]. These studies indicate that the need for dietary protein may be high at the initiation of training.

Strength Training

Frontera et al. [44] demonstrated that older (age 60–72 years) sedentary men have the capacity to significantly increase both the size and strength of their muscles. Using a progressive resistance training (PRT)

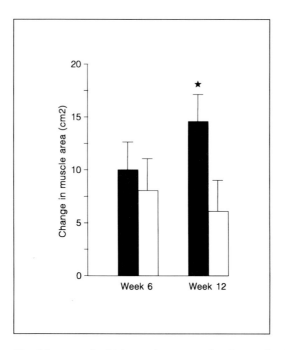

Fig. 1. Increases in thigh muscle cross-sectional area after 6 and 12 weeks of weight-lifting exercise. The dark bar represents men who drank a protein/calorie supplement and the open bar men who received no supplement. *Different from unsupplemented men. Adapted from Meredith et al. [45].

program (80% of the 1 repetition maximum, 3 days per week, 12 weeks), we demonstrated that muscle hypertrophy was associated with a significant post-training elevation in urinary 3-methylhistidine/creatinine. This PRT program had a substantial eccentric component, which almost certainly resulted in significant damage in the knee extensor and flexor muscles. Half of the men who participated in this study were given a daily protein/calorie supplement (S) providing an extra 560 ± 16 kcal/day (16.6% as protein, 43.8% as carbohydrate, and 40.1% as fat) in addition to their normal ad libitum diet. The rest of the subjects received no supplement (NS) and consumed an ad lib diet. By the twelfth week of the study, dietary calorie (2,960 ± 230 in S vs. 1,620 ± 80 kcal in NS) and protein (118 ± 10 in S vs. 72 ± 11 g/day in NS) intake was significantly different between the S and NS groups. Composition of the midthigh was estimated by computerized tomography and showed that the S group had greater gains in muscle than did the NS men (fig. 1). In addition, urinary creatinine excretion was greater at the end of the training in the S group when compared to that of the men in the NS group [45], indicating a greater muscle mass in the S group. The change in energy and protein intake (beginning vs. 12 weeks) was correlated with the change in mid-thigh muscle area (r = 0.69, p = 0.019; r = 0.63, p = 0.039, respectively). There was no difference in strength gains between the two groups. These data suggest that a change in total food intake, or perhaps, selected nutrients, in subjects beginning a strength training program can affect muscle hypertrophy.

It is clear that exercise-induced muscle damage leads to a long-term increase in protein breakdown and synthesis [43, 44, 46, 47]. Few studies have compared the longitudinal effects of high-intensity eccentric and concentric exercise training. Most progressive resistance training devices and regular weights have substantial concentric and eccentric components. Komi and Buskirk [48] measured arm circumference before and after training either eccentrically or concentrically. They found that arm circumference increased only in the arms of men who trained eccentrically. Ciriello et al. [49] examined the effects of 4 months of high-intensity strength training on a Cybex isokinetic dynamometer which has little or no eccentric component. Although the strength of the subjects increased significantly, there was no evidence of hypertrophy of type I or type II muscle fibers. Pearson and Costill [50] examined the effects of a progressive resistance weight-training protocol (with eccentric and concentric components) on one leg and isokinetic training (with no eccentric component) on the contralateral leg. Only the leg trained with a significant eccentric component increased in size as a result of the training. Recently, a direct comparison of the effects of strength training with eccentric and concentric or concentric exercise alone was made [51, 52]. These studies showed that increases in strength were greater following a program of maximum concentric and eccentric muscle actions than resistance training using concentric muscle actions only. The evidence suggests that eccentric exercise-induced skeletal muscle damage and its subsequent repair are important for increasing muscle fiber size in response to strength training. Not only can resistance training increase muscle size, but a recent report indicated that long-term resistance training may prevent age-associated changes in histochemical fibre-type distribution, myosin heavy chain isoforms and tropomyosin isoforms [53, 54].

Conclusions

Prolonged submaximal exercise increases the oxidation of indispensable amino acids which almost certainly increases the dietary requirement for protein. For this reason, 1.2 g·kg⁻¹·day⁻¹(twice the standard deviation seen by Meredith et al. [19]) should serve as a recommended dietary protein intake for endurance athletes consuming a eucaloric diet. Those who attempt to lose weight by increasing their activity levels and decreasing their energy intake will have an even greater dietary protein requirement. In previously sedentary individuals, the initiation of a weight-lifting program may also increase the need for dietary protein by increasing the rate of skeletal muscle protein. However, there is no evidence that strength and power athletes already adapted to their sport and who perform little or no endurance exercise have a high dietary protein requirement.

References

1 von Liebig J: Die Quelle der Muskelkraft. Ann Chem Pharmac 1870;153:157–228.
2 Fick A, Wisclicenus J: On the origin of muscular power. Philosoph Mag J Sci 1866; 31:485–503.
3 Cathcart EP: The influence of muscle work on protein metabolism. Physiol Rev 1925;5:225–243.
4 Lemon PWR, Mullin J: Effect of initial muscle glycogen levels on protein catabolism during exercise. J Appl Physiol 1980;48:624–629.
5 Fink WJ, Costill DL, Van Handel WJ: Leg muscle metabolism during exercise in the heat and cold. Eur J Appl Physiol 1975;34:183–190.
6 Tarnopolsky LJ, MacDougall JD, Atkinson SA, et al: Gender differences in substrate for endurance exercise. J Appl Physiol 1990;68:302–308.
7 Knapick J, Meredith C, Jones B, et al: Leucine metabolism during fasting and exercise. J Appl Physiol 1991;70:43–47.
8 Millward DJ, Davies CTM, Halliday D, et al: Effect of exercise on protein metabolism in humans as explored with stable isotopes. Fed Proc 1982;41:2686–2691.
9 Rennie MJ, Halliday D, Davies CTM, et al: Exercise induced increase in leucine oxidation in man and the effect of glucose; in Walser, Williamson (eds): Metabolism and Clinical Implications of Branched Chain Amino and Ketoacids. New York, Elsevier/North Holland, 1980, p 361–366.
10 Rennie MJ, Edwards RHT, Krywawych S, et al: Effect of exercise on protein turnover in man. Clin Sci Lond 1981;61:627–639.
11 Fielding RA, Evans WJ, Hughes VA, et al: The effects of high intensity exercise on muscle and plasma levels of alpha-ketoisocaproic acid. Eur J Appl Physiol 1986; 1986:482–485.

12 Devlin JT, Brodsky I, Scrimgeour A, et al: Amino acid metabolism after intense exercise. Am J Physiol 1990;258:E249–E255.

13 Carraro F, Stuart CA, Hartl WH, et al: Effect of exercise and recovery on muscle protein synthesis in human subjects. Am J Physiol 1990;259:E470–E476.

14 Lemon PWR, Dolny DG, Yaresheski KE: Effect of intensity on protein utilization during prolonged exercise. Med Sci Sports Exerc 1984;16:151.

15 Haralambie G, Berg A: Serum urea and amino nitrogen changes with exercise duration. Eur J Appl Physiol Occup Physiol 1976;36:39–48.

16 Gontzea I, Sutzescu P, Dumitrache S: The influence of muscle activity on nitrogen balance and the need of man for proteins. Nutr Rep Int 1974;10:35–43.

17 Gontzea I, Sutzescu R, Dumitrache S: The influence of adaptation to physical effort on nitrogen balance in man. Nutr Rep Int 1975;11:231–236.

18 Tarnopolsky MA, MacDougall JD, Atkinson SA: Influence of protein intake and training status on nitrogen balance and lean body mass. J Appl Physiol 1988;64: 187–193.

19 Meredith CN, Zackin MJ, Frontera WR, et al: Dietary protein requirements and body protein metabolism in endurance-trained men. J Appl Physiol 1989;66:2850–2856.

20 Nelson ME, Fisher EC, Catsos PD, et al: Diet and bone status in amenorrheic runners. Am J Clin Nutr 1986;43:910–916.

21 Fisher EC, Nelson ME, Frontera WR, et al: Bone mineral content and levels of gonadotropins and estrogens in amenorrheic running women. J Clin Endocrinol Metab 1986;92:1232–1236.

22 Asmussen E: Observations on experimental muscular soreness. Acta Rheum Scand 1956;2:109–116.

23 Newham DJ, McPhail G, Mills KR, Edwards RHT: Ultrastructual changes after concentric and eccentric contractions of human muscle. J Neurol Sci 1983;61:109–122.

24 O'Reilly KP, Warhol MJ, Fielding RA, et al: Eccentric exercise-induced muscle damage impairs muscle glycogen repletion. J Appl Physiol 1987;63:252–256.

25 Evans WJ, Meredith CN, Cannon JG, et al: Metabolic changes following eccentric exercise in trained and untrained men. J Appl Physiol 1986;61:1864–1868.

26 Hikida RS, Staron RS, Hagerman FC, et al: Muscle fiber necrosis associated with human marathon runners. J Neurol Sci 1983;59:185–203.

27 Warhol MJ, Siegel AJ, Evans WJ, et al: Skeletal muscle injury and repair in marathon runners after competition. Am J Pathol 1985;118:331–339.

28 Friden J, Seger J, Sjostrom M, et al: Adaptive response in human skeletal muscle subjected to prolonged eccentric training. Int J Sports Med 1983;4:177–183.

29 Kampschmidt R: Leukocytic endogenous mediator/endogenous pyrogen; in Powanda PGC (ed): The Physiologic and Metabolic Responses of the Host. Amsterdam, Elsevier/North Holland, 1981, pp 55–74.

30 Cannon JG, Orencole SF, Fielding RA, et al: Acute phase response in exercise: interaction of age and vitamin E on neutrophils and muscle enzyme release. Am J Physiol 1990;259:R1214–R1219.

31 Babior BM, Kipnes RS, Curnutte JT: The production by leukocytes of superoxide, a potential bactericidal agent. J Clin Invest 1973;52:741–744.

32   Smith JK, Grisham MB, Granger DN, Korthuis RJ: Free radical defense mecha-
       nisms and neutrophil infiltration in postischemic skeletal muscle. Am J Physiol
       1989;256:H789–H793.
33   Bainton DF: Phagocytic cells: Developmental biology of neutrophils and eosino-
       phils; in Gallin JI, Goldstein IM, Snyderman R (eds): Inflammation: Basic Princi-
       ples and Clinical Correlates. New York, Raven Press, 1988, pp 265–280.
34   Johnston RB: Monocytes and macrophages. N Engl J Med 1988;318:747–752.
35   Jones DA, Newham DJ, Round JM, Tolfree SEJ: Experimental human muscle dam-
       age: Morphological changes in relation to other indices of damage. J Physiol (Lond)
       1986;375:435–448.
36   Round JM, Jones DA, Cambridge G: Cellular infiltrates in human skeletal muscle:
       exercise induced damage as a model for inflammatory muscle disease? J Neurol Sci
       1987;82:1–11.
37   Cannon JG, Kluger MJ: Endogenous pyrogen activity in human plasma after exer-
       cise. Science 1983;220:617–619.
38   Nawabi MD, Block KP, Chakrabarti MC, Buse MG: Administration of endotoxin,
       tumor necrosis factor, or interleukin 1 to rats activates skeletal muscle branched-
       chain α-keto acid dehydrogenase. J Clin Invest 1990;85:256–263.
39   Evans WJ, Fisher EC, Hoerr RA, Young VR: Protein metabolism and endurance
       exercise. Physic Sports Med 1983;11:63–72.
40   Cannon JG, Fielding RA, Fiatarone MA, et al: Interleukin–1β in human skeletal
       muscle following exercise. Am J Physiol 1989;257:R451–R455.
41   Vihko V, Salminen A, Rantamaki J: Exhaustive exercise, endurance training, and
       acid hydrolase activity in skeletal muscle. J Appl Physiol 1979;47:43–50.
42   Duan C, Delp MD, Hayes DA, Delp PD, Armstrong RB: Rat skeletal muscle mino-
       chondrial [Ca2+] and injury from downhill walking. J Appl Physiol 1990;68:1241–
       1251.
43   Fielding RA, Meredith CA, O'Reilly KP, et al: Enhanced protein breakdown follow-
       ing eccentric exercise in young and old men. J Appl Physiol 1991;71:674–679.
44   Frontera WR, Meredith CN, O'Reilly KP, Knuttgen HG, Evans WJ: Strength con-
       ditioning in older men: Skeletal muscle hypertrophy and improved function. J Appl
       Physiol 1988;64:1038–1044.
45   Meredith CN, Frontera WR, Evans WJ: Body composition in elderly men: Effect of
       dietary modification during strength training. J Am Geriatr Soc 1992;40:155–162.
46   Cannon JG, Meydani SN, Fielding RA, et al: Acute phase response in exercise. II.
       Associations between vitamin E, cytokines, and muscle proteolysis. Am J Physiol
       1991;260:R1235–R1240.
47   Evans WJ: Exercise and muscle metabolism in the elderly; in Hutchinson ML,
       Munro HN (eds): Nutrition and Aging. San Diego, Academic Press, 1986, pp 170–
       191.
48   Komi PV, Buskirk EB: Effect of eccentric and concentric muscle conditioning on
       tension and electrical activity of human muscle. Ergonomics 1972;15:417–434.
49   Ciriello VN, Holden WL, Evans WJ: The effects of two isokinetic training regimens
       on muscle strength and fiber composition; in Knuttgen HG, Vogel JA, Poortmans J
       (eds): Biochemistry of Exercise. Champaign, Human Kinetics Publishers, 1983, pp
       787–793.

50   Pearson DR, Costill DL: The effects of constant external resistance exercise and
     isokinetic exercise training on work-induced hypertrophy. J Appl Sport Sci Res
     1988;2:39–41.
51   Dudley GA, Tesch PA, Miller MA, Buchanan MD: Importance of eccentric actions
     in performance adaptions to resistance training. Aviat Space Environ Med 1991;62:
     543–550.
52   Colliander EB, Tesch PA: Effects of eccentric and concentric muscle actions in
     resistance training. Acta Physiol Scan 1990;140:31–39.
53   Klitgaard H, Mantoni M, Schiaffino S, et al: Function, morphology and protein
     expression of ageing skeletal muscle: A cross-sectional study of elderly men with
     different training backgrounds. Acta Physiol Scand 1990;140:41–54.
54   Klitgaard H, Zhou M, Schiaffino S, et al: Ageing alters the myosin heavy chain
     composition of single fibres from human skeletal muscle. Acta Physiol Scand 1990;
     140:55–62.

William J. Evans, PhD, Chief, Human Physiology Laboratory, USDA Human
Nutrition Research Center on Aging, Tufts University, 711 Washington Street,
Boston, MA 02111 (USA)

Simopoulos AP, Pavlou KN (eds): Nutrition and Fitness for Athletes.
World Rev Nutr Diet. Basel, Karger, 1993, vol 71, pp 34–60

Carbohydrate Needs of Elite Athletes

Clyde Williams[1]

Department of Physical Education, Sports Science and Recreation Management,
Loughborough University, Loughborough, Leicestershire, UK

Introduction

The consensus view of health professionals is that diet plays a central role in the development and maintenance of health. They recommend that healthy diets should be composed of carbohydrate-containing foods, such that they provide us with about 50% of our daily energy intake, fats should account for no more that 35% and protein no more than 15% [1]. Carbohydrate-containing foods such as bread, potatoes, rice, pasta, cereals, fruit and vegetables provide a ready supply of glucose, which is a substrate for energy metabolism, for the central nervous system and skeletal muscles. They also provide micronutrients and fibre which are essential for health. An additional advantage gained from eating foods high in complex carbohydrates is that their bulky nature contributes to an early sense of satiation and so helps decrease overall food intake. A shift to a diet containing the recommended amounts of complex carbohydrates will help, indirectly, decrease the intake of foods which are high in saturated fats.

The reality is, however, that people who are not actively involved in sport eat foods that provide about the same proportion of energy from fats

[1] The author wishes to acknowledge the cooperation of his colleagues, Lisa Piearce, Joanne Fallowfield and Dr. George Gaitanos in completing the studies reported in this review. Parts of this review have been published in the IOC Book of Soccer.

and carbohydrates. A recent national survey reported that British men consume, on average, 2,480 kcal a day, of which 45% is carbohydrate (272 g), 38% is from fat (102 g) and 14% is from protein (85 g). British women consume 1,750 kcal a day, of which 42% is carbohydrate (193 g), 39% is from fat (74 g) and 15% from protein (62 g) [2]. The carbohydrate intake of athletes, beyond the recommendations of the health professionals, should be dictated by the quantity and quality of their daily training. But how much carbohydrate is required for athletes in training is a difficult question to answer because there are relatively few studies on the links between carbohydrate intake, glycogen accumulation and physical performance. Therefore, in this brief review of the carbohydrate needs of elite athletes, the following issues will be considered: (a) the carbohydrate intakes of athletes undertaking heavy exercise; (b) the role of carbohydrate in energy metabolism; (c) the link between the carbohydrate content of diets and exercise capacity; (d) links between changes in carbohydrate intake and recovery from exercise, and (e) increased carbohydrate intake and training capacity.

Carbohydrate Intakes of Athletes

Although there are nutritional surveys of the general population, there are only a few equivalent surveys of sportsmen and sportswomen, hereafter called athletes. One such study of the nutritional habits of elite athletes in the Netherlands reported that Dutch male runners had daily energy intakes of about 3,177 kcal and carbohydrate intakes of approximately 415 g. The elite women runners had a daily energy intake of 2,094 kcal and a carbohydrate intake of 307 g [3]. These values are similar to those of our own nutritional survey of British male and female endurance runners [Piearce and Williams, unpubl.] and those of other researchers (tables 1, 2) [4–8]. The carbohydrate intakes of male runners appear to range between 373 g (4.6 g·kg^{-1} bw) and 627 g·day^{-1} (9 g·kg^{-1} bw), whereas the values for women range from about 290 g (5.3 g·kg^{-1} bw) to 428 g·day^{-1} (7 g·kg^{-1} bw).

The quality of the results obtained are, however, only as good as the co-operation of the athletes recording their food intakes. There is always a concern that too many demands on the time of subjects taking part in nutritional surveys will lead to a decrease in their co-operation. This, and the inconvenience of weighing all food eaten for more than 3 or 4 days

Table 1. Examples of daily intakes of energy, protein, fat and carbohydrate (CHO) of well-trained male athletes

Group	Energy intake kcal	Protein g	Protein %	Fat g	Fat %	CHO g	CHO %	References
Distance runners (n = 50)	3,170	114	14	116	33	417	53	Piearce and Williams [unpubl.]
Distance runners (n = 10)	3,034	128	17	115	34	396	49	Grandjean [6]
Triathletes (n = 25)	4,095	134	13	127	27	627	60	Burke et al. [8]
Marathon runners (n = 19)	3,570	128	15	128	32	487	52	Burke et al. [8]
Football players (n = 56)	3,395	126	15	141	38	373	44	Burke et al. [8]
Weight lifters (n = 19)	3,640	156	18	155	39	399	43	Burke et al. [8]
Soccer players (n = 8)	4,952	170	14	217	39	596	47	Jacobs et al. [52]
Swimmers (n = 22)	5,222	166	12	248	43	596	45	Berning et al. [4]
Swimmers, male (n = 9)	3,072	108	15	102	30	404	55	Hawley and Williams [7]

Table 2. Examples of daily intakes of energy, protein, fat and carbohydrate (CHO) of well-trained female athletes

Group	Energy intake kcal	Protein g	Protein %	Fat g	Fat %	CHO g	CHO %	References
Distance runners (n = 44)	1,931	70	19	60	28	290	53	Piearce and Williams [unpubl.]
Distance runners, eumenorrheic (n = 33)	2,489	81	12	97	35	352	53	Deuster et al. [5]
Distance runners, amenorrheic (n = 12)	2,151	74	13	67	27	344	60	Deuster et al. [5]
Swimmers (n = 21)	3,573	107	12	164	41	428	48	Berning et al. [4]
Swimmers, females (n = 11)	2,130	79	16	63	28	292	56	Hawley and Williams [7]

are the common causes of under reporting food intake during nutritional surveys. Black et al. [9] have suggested that one way of checking the authenticity of energy intake values is to compare them to minimum requirements for energy balance. They recommended the calculation of the basal metabolic rates (BMR) of subjects, from their heights and weights [10]. Then, assuming that the minimum energy intake, required by sedentary people, to maintain energy balance is at least 1.5 times their BMR, the reported energy intakes can be easily compared with the predicted minimum values for energy expenditure. Any results below the minimum value should be treated with caution and re-examined, especially in the presence of stable body weights. This approach to the results of energy intake studies on elite male athletes generally confirms that intakes are greater than the calculated requirements for energy balance. In contrast, however, the results of studies on the food intake of female athletes tend to be lower than what would be predicted from their energy expenditures. For example, the ratio of the energy intake to calculated BMR for female runners [Piearce and Williams, unpubl.] (table 2) was 1.48, whereas the equivalent values for female swimmers are 1.49 [7] and 2.64 [4]. The latter value is well within the range of acceptability, whereas the former values for runners and one group of swimmers is below what might be expected for athletes in training. The values calculated for the eumenorrheic and amenorrheic runners (table 2) [5] were 1.98 and 1.77, respectively, suggesting that the energy intakes of these athletes met their energy needs.

Mulligan and Butterfield [11] examined the energy balance of two groups of female runners. One group consisted of moderately active runners, based on their weekly training distances, whereas the other group was made up of very active runners. The daily energy intake of the moderately active runners was 1,988 kcal and the value for the very active group was 1,973 kcal. Age-matched female nonrunners had slightly lower energy intakes than the runners, namely 1,744 kcal. The daily activities of the subjects in this study were carefully observed and recorded along with their daily food intakes. When this information was then used to calculate the energy balance of the runners, it showed a daily energy deficit of 650 kcal.

In non-weight-bearing sports such as swimming, the link between body weight and performance is not as strong as it is in, for example, running. Therefore, one might expect swimmers to have energy intakes which more closely resemble their energy intakes than those of female runners. This appears to be true for the energy intakes of the swimmers reported by Berning et al. [4] (table 2). The energy and carbohydrate intakes of these swimmers are as high as those reported for male runners (table 2). But Hawley and Williams [7] showed that not all female swimmers have high energy intakes (table 2). They also estimated the energy expenditures of their swimmers and calculated that the women had a daily energy deficit of approximately 800 kcal.

It has been suggested that the low energy intake values, commonly reported for women, are not simply a consequence of underreporting but a metabolic adaptation to large energy expenditures and reduced food intake [11]. However, in their study on energy balance Mulligan and Butterfield [11] did not find a reduction in BMR, nor a reduction in energy expenditure during the daily round of activities undertaken by the runners. In the absence of such supporting evidence for a metabolic adaptation, the only conclusion is that the women were underreporting their food intake or, more likely, eating less during the period when food intake was recorded. Therefore, their reported carbohydrate intakes were probably less than they needed for people involved in such prolonged training.

There is some evidence to suggest that when sedentary people undertake prolonged endurance training, their choice of foods changes, but it appears that only men increase their overall daily energy intake. Janssen et al. [12] found that during 18 months of endurance training the men in their study increased their energy intake, mainly by eating extra carbohydrate. Their carbohydrate intake increased from 280 g at the beginning to 346 g per day at the end of the study. The women in their study also increased their carbohydrate intake from 239 to 283 g per day but they did not increase their overall energy intake.

The daily carbohydrate intake of athletes is not only achieved at meal times. An equal amount of carbohydrates is consumed, in the form of snacks, between meals [13]. This is an important eating strategy for athletes because it allows them to achieve high carbohydrate intakes without the discomfort which occurs from eating excessively large meals. Therefore, a decrease in snacks would significantly reduce daily energy intake. Whether or not female athletes eat less or underreport, e.g. their food intake between meals, remains to be established. Answers to this question may be provided when the results of more studies on energy balance become available. The doubly labelled water technique for measuring energy expenditure provides a real opportunity for answering many of these questions on human energy balance [14].

Carbohydrate Stores

Carbohydrate is stored in muscles and in the liver as glycogen, a polymer of glucose. The liver glycogen concentration of a well-fed adult man is about 250 glucosyl units·kg^{-1} ww [15]. Liver glycogen helps maintain optimum blood glucose concentration, ensuring an adequate substrate supply for energy metabolism within the central nervous system. This source of blood glucose decreases during fasting but probably not at the rate that was reported as the result of biopsy studies [16]. More recent studies using NMR suggest that the rate of glucose provision by gluconeogenesis, in the liver, is greater than was formerly reported [17].

The concentration of glycogen in skeletal muscles varies with the carbohydrate content of the diet, the training status of the individual and, to a lesser extent, the type of muscle sampled. Muscle glycogen concentrations of untrained, well-fed sedentary individuals range from 70 to 110 mmol glucosyl units·kg^{-1} ww, whereas in well-fed and rested endurance athletes the values range from 140 to over 230 mmol glucosyl units·kg^{-1} ww [18]. Freeze drying muscle samples, before analyzing them for glycogen and glycolytic intermediates, increases the sensitivity of the analytical techniques. It removes the water in muscle and, as a consequence, the concentrations of metabolites increase proportionally. Therefore, results of analyses of freeze-dried muscle are expressed as dry weight (dw), rather than wet weight. Assuming an average water content of human skeletal muscle of about 77% then wet weight concentrations for muscle metabolites and substrates increase by a factor of 4.3 when expressed as dry weight values. This conversion factor is useful because not all the glycogen concentrations reported in the literature are expressed in terms of wet weight muscle.

Muscle glycogen stores are reduced during prolonged heavy exercise but they can be restocked to values which are greater than pre-exercise concentrations by consuming a diet which is rich in carbohydrates. This dietary carbohydrate loading after prolonged heavy exercise produces a phenomenon called glycogen supercompensation. This phenomenon was first reported by Bergstrom and Hultman [19]. They used a modified form of the original Duchenne biopsy needle to obtain repeated samples of muscle to follow the changes in glycogen resynthesis after exercise. The experimental design was quite novel because it involved single leg exercise with the contralateral leg acting as a resting control. Bergstrom and Hultman were their own subjects and they contributed one exercise leg and one control leg. They exercised to exhaustion using a cycle ergometer placed

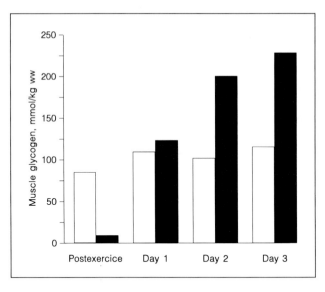

Fig. 1. Muscle glycogen concentrations in control and exercised legs during recovery. After Bergstrom and Hultman [19].

between them. Muscle samples from the vastus lateralii were obtained from the exercise and the control legs before, immediately after exercise and again at 24-hour intervals for 3 days. During the 3 days following the experiment, the 2 subjects consumed a diet which was reported as 'consisting almost exclusively of carbohydrate'. The energy intake during the remainder of the experimental day and the following 2 days was reported as 2,200 and 2,600 kcal for the 2 subjects, respectively. Assuming that their diet did consist almost entirely of carbohydrate, as reported, then the 2 subjects consumed approximately 550 and 650 g of carbohydrate a day. The muscle glycogen concentration in the exercised legs increased to values which were twice as high as glycogen concentration in the control legs during the 3-day recovery period (fig. 1).

It is important to recognize, however, that the glycogen supercompensation phenomenon occurs only in previously active muscle and so it is a local phenomenon. Further exploration of this interesting phenomenon has shown an inverse relationship between the activity of the active form of the enzyme, glycogen (EC 2.4.1.11) synthase and the concentration of glycogen in muscle [13, 20, 21]. The specificity of this relationship has been highlighted in a study which examined, both biochemically and histochemically, the activity of this enzyme in samples of different populations of muscle fibres, obtained from subjects who had exercised to exhaustion. The activity of glycogen synthase was greatest in the type I fibres, i.e. the slow contracting oxidative fibres [22]. From more recent work it can be concluded that this population of fibres is most involved during submaximal cycling to exhaustion [23]. Trained individuals, irrespective of their sport, appear to have higher resting muscle glycogen concentrations than untrained people [24, 25]. More specifically, when only one limb has been trained, it and not the resting limb has the highest glycogen concentration [22].

While the cabohydrate loading procedure is well known, it is often overlooked that this dietary manipulation also increases the glycogen stores in the liver [15]. Blood glucose, derived from liver glycogen, contributes to carbohydrate metabolism in working muscles, especially towards the end of prolonged exercise when muscle glycogen concentrations are low [27–29]. However, the relative contribution of liver glycogen to improved endurance performance is difficult to separate out from the influence of increased muscle glycogen stores.

The traditional method of carbohydrate loading involves exercise to exhaustion, to lower muscle glycogen concentrations in skeletal muscles, followed by 3 days on a low carbohydrate diet and then a high carbohydrate diet for about 3 days before endurance competitions [30]. However, athletes find the 3 days on a low carbohydrate diet unpleasant and often intolerable, especially if they continue to train hard. Sherman et al. [25] showed that the low carbohydrate phase of the traditional carbohydrate loading procedure can be left out without sacrificing glycogen supercompensation. They demonstrated that a carbohydrate-rich diet, consumed for 3 days prior to competition, along with a gradual reduction in training volume, results in increased muscle glycogen concentrations of the same magnitude as those achieved with the traditional carbohydrate loading procedure.

Studies on the links between an increase in carbohydrate intake, muscle glycogen and exercise capacity, limited in number as they are, can be divided into two groups. There are those studies that have reported the influence of dietary carbohydrate loading on endurance capacity and those which have reported the influence of this dietary manipulation on endurance performance. Endurance capacity is the time to exhaustion during exercise of constant intensity, whereas endurance performance is the time taken to complete a prescribed workload or distance. Improvements in endurance capacity, after carbohydrate loading, are greater than the improvements in endurance performance. The following sections provide a description of some of the principal studies on the influence of an increased carbohydrate intake on exercise capacity. They have been divided into those studies which have used cycling as the mode of exercise and those which have used running. The reason for considering these two common forms of exercise separately is that there is some evidence to suggest that carbohydrate loading may not produce the same improvements in exercise performance during running as it does during cycling.

Carbohydrate Intake and Endurance Capacity (Cycling):
Energy Metabolism during Submaximal Exercise

During prolonged exercise of submaximal intensity, the aerobic metabolism of fatty acids and glycogen are the main substrates for ATP resynthesis. The relative contributions of fat and carbohydrate to energy metabolism in working muscles depends on the training status of the individual and also the relative exercise intensity. Relative exercise intensity is the oxygen cost of the activity expressed as a percentage of the athlete's maximal oxygen uptake ($\%VO_2$ max). Training increases the aerobic capacity of skeletal muscles, mainly by an increase in mitochondrial density and in the number of capillaries around type I fibres [31]. The type I muscle fibres are slow contracting and slow to fatigue. They rely on aerobic metabolism of fatty acids and carbohydrate to regenerate ATP from ADP [31]. When muscle glycogen concentration falls to a critically low value, the rate of aerobic metabolism decreases because of the concomitant decrease in mitochondrial Krebs cycle intermediates. Although there is an abundance of fatty acids available for muscle metabolism, they cannot undergo oxidation fast enough to provide ATP at the rate at which it is used by working muscle [32]. The result is a decrease in contractile activity which is manifest as whole body fatigue.

Aerobic metabolism of fatty acids and carbohydrate is the most economical use of these two substrates, this is particularly true for the limited carbohydrate stores. However, partial degradation of muscle glycogen to lactate provides 3 mmol of ATP without the involvement of oxidative metabolism. This anaerobic provision of ATP has the advantage that it can contribute rapidly to the energy needs of working muscles when aerobic

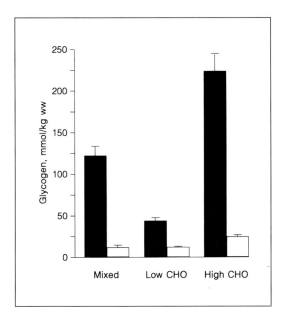

Fig. 2. Pre- and postexercise muscle glycogen concentrations after three dietary conditions. After Bergstrom et al. [34].

metabolism cannot cope with the increased demands placed on this system [32].

The early studies of Christensen and Hansen [33] were the first to firmly establish the link between a high carbohydrate diet and the improvement in endurance capacity. In these studies the endurance capacity of a group of subjects was examined, on a cycle ergometer, after 3–4 days on either a normal mixed diet, a diet of fat and protein or a diet rich in carbohydrate. After the high carbohydrate diet the endurance capacity of these subjects doubled in comparison with their exercise times after they had been consuming their normal mixed diets. In contrast, the fat and protein diet reduced the exercise performance to almost half that achieved on a normal mixed diet. Bergstrom et al. [34] examined the influence of high pre-exercise muscle glycogen concentrations on endurance capacity (fig. 2). They found that the carbohydrate loading procedure increased the endurance capacity of their subjects by over 50% compared with the times they achieved when they ate their normal diets (fig. 3).

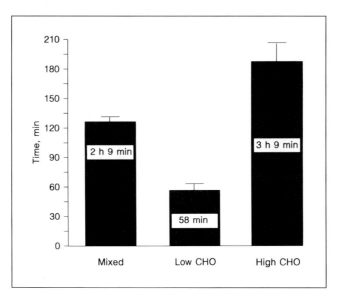

Fig. 3. Exercise times to exhaustion after three dietary conditions. After Bergstrom et al. [34].

Carbohydrate Intake and Endurance Capacity (Running)

Although carbohydrate loading was readily adopted by runners in preparation for endurance competitions, the evidence supporting this practice is based almost entirely on cycling studies. There are relatively few studies showing that carbohydrate loading improves endurance capacity during running. In one study in which the influence of the traditional carbohydrate loading procedure on endurance running capacity was examined, Goforth et al. [35] showed that endurance time was increased from 120 to 130 min. The treadmill running speeds for the 9 well-trained runners in their study was equivalent to 80% VO_2 max. The dietary manipulation during the carbohydrate loading procedure was well controlled. Isocaloricity of each phase of the carbohydrate loading was achieved using a combination of liquid and solid foods. The daily energy intake of the subjects was 3,500 kcal, of which 46% was from carbohydrate (429 g), 42% was from fat and 12% was from protein. During the high carbohydrate phase of the diet, the relative contributions of carbohydrate, fat and protein to the daily energy intake were 64 (597 g), 24 and 12%, respectively.

The carbohydrate intake of their control diet was quite high and so this, along with the high relative exercise intensity (%VO$_2$ max) may explain why there was an increase of only 9% in endurance capacity. This increase in capacity was quite modest in comparison with the improvements of 57% reported in the cycle ergometer studies of the late 1960s [34].

A clear benefit from increasing the carbohydrate content of runners' diets was reported by Brewer et al. [36]. In their study, the normal mixed diets of 30 runners were supplemented with either additional protein, or complex carbohydrates or simple carbohydrates. The 3 groups of subjects ran to exhaustion on a level treadmill at speeds equivalent to 70% VO$_2$ max before and after 3 days of dietary supplementation. Running times, after the high carbohydrate diets, increased by 26 and 23% for the complex carbohydrate group and the simple carbohydrate group, respectively. But there was no improvement in the performance times of the control group after supplementing their normal diets with additional protein and fat. Previous studies have shown that supplementing normal diets with simple carbohydrates, such as confectionery products, is as effective in restoring muscle glycogen as are complex carbohydrates [37]. This means that athletes who find that the bulky nature of complex carbohydrate foods limits the amount they can eat can consume additional carbohydrate in the form of glucose drinks and confectionery products as part of the carbohydrate loading procedure.

In contrast, a more recent study showed no improvement in endurance running capacity after carbohydrate loading. Madsen et al. [26] reported that carbohydrate loading increased muscle glycogen concentrations in the gastrocnemius muscles of their subjects by 25% (581–772 mmol·glucosyl units kg^{-1} dw). Their subjects ran to exhaustion on a level treadmill at speed equivalent to 75–80% VO$_2$ max on two occasions. The time to exhaustion on the first trial, after the subjects consumed their normal mixed diet was 70 min, whereas after carbohydrate loading, using the method described by Sherman et al. [25], they recorded 77 min. This 10% improvement in endurance capacity was not, however, statistically significant. The amount of muscle glycogen used was the same in both trials. Furthermore, the average glycogen concentrations at exhaustion, after the control and carbohydrate trials, were 553 and 434 mmol·glucosyl units·kg^{-1} dw, respectively. These values do not reach the critically low levels associated with fatigue during cycling at the same relative exercise intensity. Histochemical examination of the samples of muscle from the runners did not show glycogen depletion in type I or type II fibres. Therefore glycogen depletion was not the cause of fatigue in these well-trained runners. Their running times were, however, considerably less than what would be expected from well-trained runners. Perhaps the biopsies and indwelling catheters prevented the runners from performing at their best under these conditions. Nevertheless, the study confirms earlier observations that running does not reduce muscle glycogen concentrations to the same low levels as does cycling to exhaustion. Therefore, there are other causes of fatigue during submaximal constant speed running which may be more important than the reduction in muscle glycogen concentration.

Carbohydrate Intake and Endurance Performance (Running)

The benefits of the carbohydrate loading in preparation for an endurance race were first reported by Karlsson and Saltin [38]. They addressed themselves to the question of whether or not an increased pre-exercise glycogen concentration improves running speed as well as endurance capacity. The study was conducted under race conditions with 2 groups of subjects. One group underwent the traditional carbohydrate loading procedure prior to an annual 30 km cross-country race, while the other group remained on their normal mixed diet. In the second part of the study the race conditions were recreated three weeks later, when the dietary preparation of each group was reversed. The time to complete the 30 km race was improved by 8 min (135.0 vs. 143.0 min) after the subjects increased their pre-race muscle glycogen concentrations by carbohydrate loading. The running speed of each subject was not increased, however, during the early part of the race, but they were able to sustain their optimum pace longer. The carbohydrate loading also resulted in postexercise glycogen concentrations that were as high as the preexercise values recorded for the subjects on their normal mixed diets (fig. 4). Therefore, the dietary preparation for the race provided glycogen stores which were more than adequate for muscle metabolism under these conditions.

In a more recent study, the influence of carbohydrate loading on running performance over 30 km was conducted using a laboratory treadmill which was instrumented so that the subjects controlled their own speeds using a light weight hand held switch [39]. Changes in speed, time and distance elapsed were all displayed on a computer screen in full view of the subjects. Although there was no overall improvement in performance times of the 18 runners (12 males and 6 females), running speeds over the

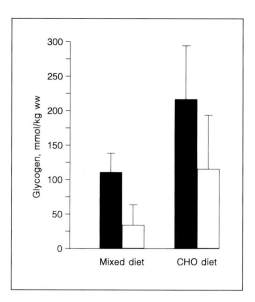

Fig. 4. Pre- and postrace muscle glycogen concentrations after mixed and high carbohydrate diets. After Karlsson and Saltin [38].

last 5 km of the simulated races were significantly faster after carbohydrate loading. Furthermore, the men improved their running times by 2.8% (127.4 vs. 131.0 min) after carbohydrate loading. The control group consumed additional protein and fat to match the energy intake of the carbohydrate group but the intake of extra food, low in carbohydrate, did not produce an improvement in endurance performance. The carbohydrate group not only ran faster during the latter part of the race than the control group but they also had lower adrenaline concentrations. The results of this simulated race over 30 km supports the conclusions reported by Karlsson and Saltin [38] that running performance is improved after carbohydrate loading before long distance races.

In contrast, Sherman et al. [25] found no differences in performance when a group of 6 well-trained endurance athletes completed 3 races over 20.9 km, after carbohydrate loading before 2 of the races. Three different dietary procedures were used to prepare for the races, namely a low carbohydrate diet followed by 3–4 days on a high carbohydrate diet (low/high; 104 vs. 542 g of carbohydrate), normal mixed diet followed by the same period on a high carbohydrate diet (mixed/high; 352 vs. 542 g of carbohy-

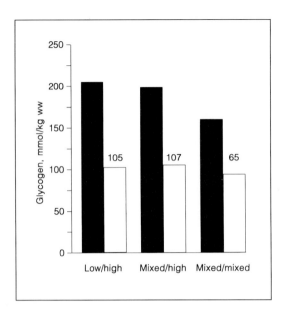

Fig. 5. Pre- and postrace muscle glycogen concentrations after three dietary conditions. After Sherman et al. [25].

drate) and a normal mixed diet for the whole of the preparatory period before the race (mixed/mixed; 353 g of carbohydrate). The running times for the 3 races were 83.5 min (low/mixed), 83.63 min (mixed/high) and 82.95 min (mixed/mixed). There were no significant differences between the performance times for the three races. Both the carbohydrate loading procedures (low/high and mixed/high) increased muscle glycogen concentrations in the gastrocnemius muscles of the runners prior to the races. The trial in which the subjects consumed their normal mixed diet also increased their muscle glycogen concentration prior to the race but not to the same extent as in carbohydrate loading trials (fig. 5). The increase in the glycogen stores without deliberate carbohydrate loading is probably a consequence of the decreased training which allows muscle time to restore its normal high glycogen concentrations.

The pre-race muscle glycogen concentrations, even without carbohydrate loading, were high, and higher than the values reported by Karlsson and Saltin [38] for their subjects before carbohydrate loading. Therefore, the muscle glycogen concentrations of the subjects in the study reported by

Sherman et al. [25] were more than sufficient to meet the demands imposed upon them by races over 20.9 km. It is clear from this study, however, that well-trained runners need only taper their training in preparation for races over the half marathon distance. They do not need to undertake any dietary manipulation in preparation for races over this or shorter distances. Furthermore, the study showed that the greater the preexercise muscle glycogen concentrations the more glycogen was used during the race (fig. 5). This is an interesting phenomenon, which appears to be metabolically wasteful, especially in light of the evidence which shows that there were no differences in running speeds between the trials.

This study by Sherman et al. [25] confirmed three important observations which have been reported separately in other studies. The first is that high glycogen concentrations can be achieved without undergoing the low carbohydrate phase of the traditional carbohydrate loading procedure, the second is that carbohydrate loading after running to exhaustion does produce glycogen supercompensation as effectively as has been repeatedly shown in cycling studies, and, third, more glycogen is used by glycogen loaded muscle than in muscle with its normal complement of carbohydrate.

Carbohydrate Intake and Performance (Multiple-Sprint Sports): Exercise of Maximal Intensity (Sprinting)

During heavy exercise lasting only 4–6 min (~ 90–100% VO_2 max) the type II fibres lose significant amounts of glycogen. So much so that the loss of this limited carbohydrate store contributes to fatigue [40]. The early studies reporting the reduction of glycogen in type II fibres were based on histochemical rather than a direct biochemical analysis [40, 41]. However, more recent studies have confirmed the greater glycogen reduction in type II fibres during high-intensity exercise using single fibre analysis [42].

Carbohydrate loading has been reported to improve performance during high, but not maximal, intensity exercise. Maughan and Poole [43] reported a 36% improvement in their subjects' cycling time to exhaustion (6.65 vs. 4.87 min), at 105% VO_2 max, after undergoing the traditional carbohydrate loading procedure. In subsequent studies the authors provided evidence to suggest that the improvement in their subjects' perfor-

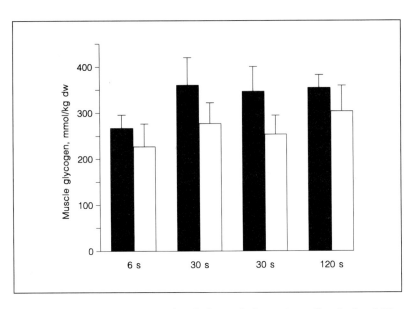

Fig. 6. Muscle glycogen concentrations before and after sprint cycling for 6 and 30 s, 30 s sprint running and running for 120 s. After Boobis [46].

mance was the consequence of an increased buffering capacity rather than as a result of the additional glycogen stores [44]. Fatigue during high-intensity exercise is associated with, but not the direct consequence of, an increase in hydrogen ion concentration in skeletal muscles and so an increased buffering capacity would be an advantage during high-intensity exercise [45].

Brief periods of exercise of maximal intensity (<30 s) demand high rates of energy production. Under these conditions ATP resynthesis is almost entirely dependent upon the contributions of phosphocreatine (PCr) and anaerobic glycogenolysis. For example, during 6 s of maximal exercise on a cycle ergometer, PCr and anaerobic glycogenolysis contributed equally to the high rate of ATP production [45]. Glycogen availability is not a limiting factor during a single, brief bout of exercise of maximal intensity. This is illustrated in figure 6, which shows the muscle glycogen concentrations before and immediately after 6 and 30 s of maximal exercise during cycling (using the Wingate protocol), glycogen concentrations before and after 30 s of sprint running, on a nonmotorized treadmill, and

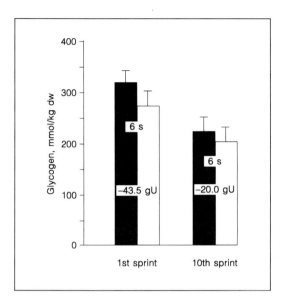

Fig. 7. Muscle glycogen concentrations before and after 10 sprints on a cycle ergometer, each sprint lasting 6 s. After Gaitanos [49].

glycogen concentrations before and after 120 s of running on a motorized treadmill, at speeds equivalent to 110% VO_2 max [46, 47]. Furthermore, increased glycogen concentrations as a result of dietary carbohydrate loading has no effect on the peak and mean power outputs during 30 s of maximal exercise on a cycle ergometer [48]. This also confirms that during one brief sprint, fatigue is not a consequence of inadequate muscle glycogen concentrations in skeletal muscles.

Repeated brief periods of maximal exercise make demands on muscle glycogen stores; however, the rate at which muscle glycogen is used decreases as exercise continues. Gaitanos [49] showed that during 10 maximal sprints of 6 s duration and 30 s recovery, on a cycle ergometer, glycogenolysis was reduced by half during the last sprint (fig. 7). There appears to be some mechanism which spares the limited glycogen stores during repeated exercise of maximal intensity. The glycogen sparing may be the result of a reduction in glycogenolysis per se and/or the consequence of an increase in the aerobic metabolism of glycogen and hence a higher energy yield per glucosyl unit. It is, nevertheless, the availability of an adequate concentration of phosphocreatine which dictates whether or not athletes can reproduce their sprint performances. An inadequate recovery between sprints will result in a failure to resynthesize sufficient phosphocreatine and so a progressive decrease in performance will occur.

Sports which involve several brief sprints do not exhaust the limited muscle glycogen stores of the participants. But sports which demand that their participants perform a combination of submaximal running and brief periods of sprinting, such as soccer, do reduce muscle glycogen concentrations to critically low values [5]. When this occurs, performance is impaired and so the assumption is that carbohydrate loading would probably be of benefit to participants in multiple-sprint sports [50]. There is some evidence to support this reasonable assumption. Bangsbo et al. [5] showed that an increased carbohydrate intake, prior to a field test which simulates the activities which are common in soccer, improves performance. Their subjects were 7 professional football players who performed two tests, after their normal mixed diet and after 2 days on a high carbohydrate diet (65% CHO, i.e. 600 g/day). The field test consisted of two parts. During the first part, the subjects performed 46 min of walking and running, simulating the activity pattern common to soccer. The second part was a performance run on a laboratory treadmill. The subjects ran at high (15 s) and low speeds (10 s) on a treadmill until exhaustion. After the high carbohydrate diet the subjects were able to run 0.9 km longer than when the test was performed after a normal mixed diet. Muscle glycogen concentrations were not determined during this particular study but it confirms the recommendations of Saltin [50] and Jacobs et al. [52] that soccer players should increase their carbohydrate intake as part of their match preparations.

Carbohydrate Intake and Recovery

In previous sections examples of studies have been cited which clearly show that heavy exercise reduces skeletal muscle glycogen concentrations to critically low levels. A diet containing between 500 and 600 g of carbohydrate a day has been recommended in order to restore muscle glycogen to pre-exercise values within 24 h [53]. In order to achieve this recommended amount, the anorexic effects of heavy exercise have to be overcome in order to ensure that athletes eat enough to provide sufficient substrate to restore muscle glycogen. The most rapid rate of glycogen resynthesis occurs during the immediate postexercise period. Ivy [54] has shown that the maximum rate of glycogen resynthesis occurs with carbohydrate intakes of between 0.8 and 3.0 g·kg^{-1} body weight, the recommended amount being about 1 g·kg^{-1} body weight, consumed every 2 h for up to 5 h of recovery. Not only is muscle glycogen resynthesized rapidly after ingesting this amount of carbohydrate but endurance capacity is greater than when no carbohydrate is consumed. For example, two groups of subjects ran to exhaustion at 70% VO$_2$ max on a level treadmill, 4 h after running for 86 min at the same relative exercise intensity. The group which ingested the recommended amount of carbohydrate [54] ran for 62 min, whereas the running time for the group which drank a sweet placebo was only 40 min [Fallowfield and Williams, unpubl.].

When recovery within 24 h is necessary, for example, during participation in tournaments that have qualifying rounds, then the normal daily carbohydrate intake of 4 to 6 g·kg^{-1} body weight may be insufficient to replace muscle glycogen and restore endurance capacity. Fallowfield and Williams [55] found that recovery from 90 min of treadmill running was incomplete when a group of experienced runners ate their normal diets without deliberately increasing their carbohydrate intake. However, a matched group of runners who increased their daily carbohydrate intake from 5.5 to 9 g·kg^{-1} bw, were able to reproduce their training performance of 90 min of treadmill running.

In contrast, Keizer et al. [56] reported that restoration of muscle glycogen was not accompanied by the capacity to reproduce performance during a maximal oxygen uptake test. The maximal work capacity of their subjects was about 8% lower after 22 h of recovery, even though muscle glycogen concentrations were back to normal. An interesting additional observation from this study was that subjects who were allowed to eat ad libitum did not replace their muscle glycogen to the same extent as those subjects whose food intake was prescribed during the first 5 h of recovery [56]. This is a particularly important observation which should be taken into account when athletes have to recover within 24 h.

Carbohydrate Intake and Training

Pascoe et al. [57] showed that an hour's training each day for 3 days lowers muscle glycogen such that on the third day the pre-exercise muscle glycogen concentrations were lower than on the first day. Even lower pre-exercise muscle glycogen concentrations were reported after 3 days of the same amount of training on a cycle ergometer. The carbohydrate intake was 5.5 g·kg^{-1}·day^{-1} and the results of the study suggest that this amount was insufficient to support daily resynthesis of muscle glycogen for several days of training. Nevertheless, these subjects were able to complete an hour's training at 75% VO$_2$max on each of the 3 days. The endurance capacities of the subjects in this experiment were not determined but daily training does not usually require athletes to work to exhaustion. Therefore, the results of this study suggest that their carbohydrate intake of 6 g· kg^{-1}·day^{-1} was almost sufficient for their needs. The duration and intensity of exercise will determine the amount of glycogen used and so if training is not heavy or prolonged and recovery does not have to be completed in 24 h then the overall daily carbohydrate intake need not necessarily be increased beyond about 6–7 g·kg^{-1}·day^{-1}.

There is some evidence that an increased carbohydrate intake enables athletes to undertake more training. Simonsen et al. [58] reported that rowers were able to complete more work during training sessions when they increased their daily carbohydrate intake from 5 to 10 g·kg^{-1} bw. After 4 weeks of heavy training, those rowers who ate a greater amount of carbohydrate were able to achieve higher power outputs during time trials over 2,500 m. Kirwan et al. [59] also investigated the links between increased carbohydrate intake and training capacity. They increased the training load of a group of well-trained runners by about 50% and pre-scribed the amount of carbohydrate needed to cover the additional train-ing. The runners were only just able to cope with the increased energy expenditure even when their carbohydrate intake was increased to 10 g·kg^{-1} bw·day^{-1}.

Costill et al. [60] reported that swimmers were able to undertake addi-tional training when they increased their carbohydrate intake to about 8 g·kg^{-1}·day^{-1}. Although these experienced competitive swimmers did cope with the additional training load, their performances in a series of swimming tests after the training period did not improve. Four of the 8 swimmers in this study failed to increase their carbohydrate intake to the required level and were unable to cope with the demands of higher training volume. In contrast, when the diets of swimmers were closely controlled so that they achieved the prescribed amount of additional carbohydrate, they were all able to cope with increased training volumes [61]. Lamb et al. [61] found that increasing the carbohydrate intake of two groups of experienced swimmers to 6.5 g·kg^{-1} bw·day^{-1} and 12.5 g·kg^{-1} bw·day^{-1}, respectively, for 9 days of increased training produced no differences in performances between the two groups. Sherman and Wimer [62] have also questioned the assumption that an increase in the daily carbohydrate intake of athletes enables them to train harder.

Performance tests that involve brief periods of exercise of maximal intensity may not be an appropriate way of examining the benefits, for training, of an increased carbohydrate intake. In these tests, rate of performing exercise is the criterion rather than exercise capacity. It is important to realize that increased glycogen stored in skeletal muscles has been consistently shown to improve endurance capacity to a greater extent than endurance performance [36, 38]. For example, brief periods of maximal exercise do not tax the glycogen stores to the same extent as continuous exercise for the same period of time. During a 5- to 6-second period of maximal exercise about half of the energy is provided by phosphocreatine. If the recovery between periods of exercise is long enough for most of the phosphocreatine to be resynthesized, then these brief periods of exercise can continue for several minutes. Similarly, Symons and Jacobs [63] showed that maximal exercise during weight lifting can be achieved when muscle glycogen concentrations are low.

It is, however, important to remember that a high carbohydrate diet increases the capacity for exercise rather than the rate of performing exercise. Exercise tests which measure, for example, peak power output or running speed are not improved by an increase in carbohydrate intake [38, 48]. Therefore, the evidence that a high carbohydrate diet increases the capacity for training remains inconclusive, because the studies are few in number and some have used inappropriate methods of measuring training. There is also another dimension to the discussion on carbohydrate intake and training which is often overlooked. It is the glycogen sparing effect of training per se. Training increases the oxidative capacity of skeletal muscles and as a consequence fat makes a greater contribution to energy metabolism during exercise than before training [64]. The more well trained athletes become, the less glycogen they will use during a standard training session [65]. Therefore, they should be able to increase their training volume at the expense of the same amount of glycogen as they used before becoming well trained. This is, of course, a testable hypothesis, especially if the training study is carried out in a laboratory where the changes in the conributions of carbohydrate to energy metabolism during exercise, after training, can be easily determined.

Conclusion

The question of how much carbohydrate should an athlete consume still cannot be answered with any great precision. A carbohydrate intake of 70% or more of the daily energy intake has been proposed [66] for athletes in training. Of course, the pitfall of recommendations based on the proportion of daily energy intakes is that the energy intake can vary among individuals. A daily carbohydrate intake of between 500 and 600 g has been proposed as being sufficient to replete skeletal muscle glycogen stores during a recovery period of 24 h [53]. This amounts to about $8.5 \, g \cdot kg^{-1}$ $bw \cdot day^{-1}$ of carbohydrate for an athlete with a body weight of 70 kg and an energy intake of about 2,400 kcal. The available evidence on carbohydrate intake, recovery and performance suggests that this amount of carbohydrate may be too little for athletes, especially those with body weights greater than 70 kg. In contrast, a recommendation of a daily carbohydrate intake of 600 g for female athletes would provide virutally all their energy intake for a day (table 2). Therefore, recommendations for carbohydrate intake based on body weights seem to have more to commend them than recommendations based on absolute amounts. Translating this information into recommendations for athletes who have little knowledge of nutrition presents another demanding challenge. Moses and Manore [67] have provided a method for monitoring carbohydrate intake in a simple and achievable way. They designed a system which allocates points for foods with different carbohydrate contents. Athletes carry a table which lists foods and their carbohydrate point scores. This practical information allows them to select foods which are not only high in carbohydrates but it also gives them the flexibility to eat a wide range of carbohydrate-containing foods. This approach is helpful and necessary because it attempts to translate nutritional principles into practice. In conclusion, the research evidence clearly shows that athletes need to eat enough carbohydrate-containing foods to ensure that their glycogen stores are restocked before exercise; however, how much muscle glycogen is 'enough' has yet to be established.

References

1   Committee on Medical Aspects of Food Policy: Diet and Cardiovascular Disease. Report of the Panel on Diet in Relation to Cardiovascular Disease. Department of Health and Social Security. Report on Health and Social Subjects, No 28. London, HMSO, 1984.

2   Gregory J, Foster K, Tyler H, Wiseman M: The Dietary and Nutritional Survey of
    British Adults. London, HMSO, 1990.
3   van Erp-Baart AMJ, Saris WHM, Binkhorst RA, Vos JA, Elvers JWH: Nationwide
    survey on nutritional habits in elite athletes. I. Carbohydrate, protein and fat intake.
    Int J Sports Med 1989;10(suppl1):S3–S10.
4   Berning JR, Troup JP, Van Handel PJ, Daniels J, Daniels N: The nutritional habits
    of adolescent swimmers. Int J Sports Nutr 1991;1:240–248.
5   Deuster PA, Kyle SB, Moser PB, Vigersky RA, Singh A, Schoomaker EB: Nutri-
    tional intakes and status of highly trained amenorrheic and eumenorrheic women
    runners. Fertil Steril 1986;46:636–643.
6   Grandjean AC: Macro-nutrients intake of US athletes compared with the general
    population and recommendations for athletes. Am J Clin Nutr 1989;49:1070–
    1076.
7   Hawley JA, Williams MM: Dietary intakes of age-group swimmers. Br J Sports Med
    1991;25:154–158.
8   Burke LM, Gollan RA, Read RSD: Dietary intakes and food use of groups of elite
    Australian male athletes. Int J Sports Nutr 1991;1:378–394.
9   Black AE, Goldberg GR, Jebb SA, Livingstone MBE, Cole TJ, Prentice AM: Critical
    evaluation of energy intake of data using fundamental principles of energy physiol-
    ogy. 2. Evaluating the results of published surveys. Eur J Clin Nutr 1991;45:583–
    599.
10 FAO/WHO/UNU: Energy and Protein Requirements. Report of a Joint Expert Con-
    sultation. Techn Rep Ser No 724. Geneva, WHO, 1985, p 71.
11 Mulligan K, Butterfield GE: Discrepancies between energy intake and expenditure
    in physically active women. Br J Nutr 1990;64:23–36.
12 Janssen GME, Graef CJJ, Saris WHM: Food intake and body composition in novice
    athletes during a training period to run a marathon. Int J Sports Med 1989;10(suppl
    1):S17–S21.
13 Kirsh KA, von Ameln H: Feeding patterns of endurance athletes. Eur J Appl Physiol
    1981;47:197–208.
14 Westerterp KR, Saris WHM: Limits of energy turnover in relation to physical per-
    formance, achievement of energy balance on a daily basis. J Sports Sci 1991;
    9(suppl):1–15.
15 Hultman E, Bergstrom J, Roch-Norland AE: Glycogen storage in human muscle; in
    Pernow B, Saltin B (eds): Muscle Metabolism during Exercise. New York, Plenum
    Press, 1971, pp 273–288.
16 Nilsson LH, Hultman E: Liver glycogen in man – the effect of total starvation or a
    carbohydrate-poor diet followed by carbohydrate refeeding. Scand J Clin Lab Invest
    1973;32:325–330.
17 Rothman DL, Magnusson I, Katz LD, Shulman RG, Shulman GI: Quantitation of
    hepatic glycogenolysis and gluconeogenesis in fasting humans with [13]C. Science
    1991;254:573–576.
18 Costill DL: Carbohydrates for ecercise: dietary demands for optimum performance.
    Int J Sports Med 1988;9:1–18.
19 Bergstrom J, Hultman E: Muscle glycogen synthesis after exercise: An enhancing
    factor localized to the muscle cells in man. Nature 1966;210:309–310.
20 Adolphsson S: Effects of contractions in vivo on glycogen content and glycogen
    synthetase activity in muscle. Acta Physiol Scand 1973;88:189–197.

21 Zachweija JJ, Costill DL, Pascoe DD, Robergs RA, Fink WJ: Influence of muscle glycogen depletion on the rate of resynthesis. Med Sci Sports Exerc 1991;23:44–48.

22 Piehl K, Adolfsson S, Nazar K: Glycogen storage and glycogen synthetase activity in trained and untrained muscle of man. Acta Physiol Scand 1974;90:779–788.

23 Vollestad NK, Vaage O, Hermansen L: Muscle glycogen depletion patterns in type I and subgroups of type II fibres during prolonged severe exercise in man. Acta Physiol Scand 1984;122:433–441.

24 Gollnick PD, Armstrong RB, Saltin B, Saubert CW, Sembrowich WL, Shepherd RE, Saltin B: Effect of training on enzyme activity and fiber composition of human skeletal muscle. J Appl Physiol 1973;34:107–111.

25 Sherman WM, Costill DL, Fink W, Miller J: Effect of exercise-diet manipulation on muscle glycogen and its subsequent utilization during performance. Int J Sports Med 1981;2:114–118.

26 Madsen K, Pedersen PK, Rose P, Richter EA: Carbohydrate supercompensation and muscle glycogen utilization during exhaustive running in highly trained athletes. Eur J Appl Physiol 1990;61:467–472.

27 Wahren J: Substrate utilization by exercise muscle in man. Prog Cardiol 1973;2:255–280.

28 Bonen A, Malcolm SA, Kilgour RD, MacIntyre KP, Belcastro AN: Glucose ingestion before and during intense exercise. J Appl Physiol 1981;50:766–771.

29 Coggan AR, Coyle EF: Metabolism and performance following carbohydrate ingestion late in exercise. Med Sci Sports Exerc 1989;21:59–65.

30 Astrand PO: Diet and athletic performance. Fed Proc 1967;26:1772–1777.

31 Saltin B, Gollnick PD: Skeletal muscle adaptability: Significance for metabolism and performance; in Peachey, Adrian, Geiger (eds): Handbook of Physiology, sect 10. Skeletal Muscle. Bethesda, American Physiological Society, 1983, pp 555–631.

32 McGilvery RW: The use of fuels for muscular work; in Howald H, Poortmans JR (eds): Metabolic Adaptation to Prolonged Physical Exercise. Basel, Birkhäuser, 1975, pp 12–30.

33 Christensen EH, Hansen O: Arbeitsfähigkeit und Ernährung. Scand Arch Physiol 1939;81:160–175.

34 Bergstrom J, Hermansen L, Hultman E, Saltin B: Diet, muscle glycogen and physical performance. Acta Physiol Scand 1967;71:140–150.

35 Goforth HW Jr, Hodgson JA, Hidlerbrand RL: A duble blind study of the effects of carbohydrate loading upon endurance performance. Med Sci Sports Exerc 1980;12:108A.

36 Brewer J, Williams C, Patton A: The influence of high carbohydrate diets on endurance running performance. Eur J Appl Physiol 1988;57:698–706.

37 Roberts KM, Noble EG, Hayden DB, Taylor AW: Simple and complex carbohydrate-rich diets and muscle glycogen content of marathon runners. Eur J Appl Physiol 1987;57:70–74.

38 Karlsson J, Saltin B: Diet, muscle glycogen and endurance performance. J Appl Physiol 1971;31:203–206.

39 Williams C, Brewer J, Walker M: The effect of a high carbohydrate diet on running performance during a 30-km treadmill time trial. Eur J Appl Physiol 1992;64:in press.

40   Essen B: Glycogen depletion of different fibre types in human skeletal muscle during intermittent and continuous exercise. Acta Physiol Scand 1978;113:446–455.

41   Friden J, Seger J, Ekblom B: Topographical localization of muscle glycogen: An ultrahistochemical study in the human vastus lateralis. Acta Physiol Scand 1989; 135:381–391.

42   Greenhaff PL, Nevill ME, Soderlund K, Bodin K, Boobis LH, Williams C, Hultman E: The metabolic responses to maximal treadmill sprinting in type I and type II human muscle fibres. J Physiol 1992;446:528P.

43   Maughan RJ, Poole DC: The effects of a glycogen loading regimen on the capacity to perform anaerobic exercise. Eur J Appl Physiol 1981;46:211–219.

44   Greenhaff PL, Gleeson M, Maughan RJ: The effects of dietary manipulation on blood acid-base status and the performance of high intensity exercise. Eur J Appl Physiol 1987;56:331–337.

45   Spriet LL, Soderlund K, Bergstrom M,. Hultman E: Skeletal muscle glycogenolysis, glycolysis, and pH during electrical stimulation in man. J Appl Physiol 1987;62: 616–621.

46   Boobis LH: Metabolic aspects of fatigue during sprinting; in Macleod D, Maughan RJ, Nimmo M, Reilly T, Williams C (eds): Exercise, Benefits, Limitations and Adaptations. London, Spon, 1987, pp 116–140.

47   Nevill ME, Boobis LH, Brooks S, Williams C: Effect of training on muscle metabolism during treadmill sprinting. J Appl Physiol 1989;67:2376–2382.

48   Wootton SA, Williams C: Influences of carbohydrate status on performance during maximal exercise. Int J Sports Med 1984;5(suppl):126–127.

49   Gaitanos GC: Human muscle metabolism during intermittent maximal exercise; unpubl PhD thesis, Loughborough University, 1990.

50   Saltin B: Metabolic fundamentals in exercise. Med Sci Sports 1973;15:366–369.

51   Bangsbo J, Norregaard L, Thorsoe F: The effect of diet on intermittent exercise performance. Int J Sports Med 1991;13:152–157.

52   Jacobs I, Weslin N, Karlsson J, Rasmussen M, Houghton B: Muscle glycogen concentration and elite soccer players. Eur J Appl Physiol 1982;48:297–302.

53   Costill DL, Sherman WM, Fink WJ, Maresh C, Whitten M, Miller JM: The role of dietary carbohydrates in muscle glycogen resynthesis after strenuous running. Am J Clin Nutr 1981;34:1813–1836.

54   Ivy JL: Muscle glycogen synthesis before and after exercise. Sports Med 1991;11: 6–19.

55   Fallowfield J, Williams C: Influence of carbohydrate feeding on recovery from prolonged exercise. Int J Sports Nutr 1992; in press.

56   Keizer HA, Kuipers H, van Kranenburg G, Geurten P: Influence of liquid and solid meals on muscle glycogen resynthesis, plasma fuel hormone response, maximal physical working capacity. Int J Sports Med 1987;8:99–104.

57   Pascoe DD, Costill DL, Robergs RA, Davis JA, Fink WJ, Pearson DR: Effect of exercise mode on muscle glycogen storage during repeated days of exercise. Med Sci Sports Exerc 1990;22:593–598.

58   Simonsen JC, Sherman WM, Lamb DR, Dernbach AR, Doyle AJ, Strauss R: Dietary carbohydrate muscle glycogen, and power output during rowing training. J Appl Physiol 1990;70:1500–1505.

59 Kirwan J, Costill DL, Mitchell JB, Houmard JA, Flynn MG, Fink WJ, Beltz JD: Carbohydrate balance in competitive runners during successive days of intense training. J Appl Physiol 1988;65:2601–2606.

60 Costill DL, Flynn MJ, Kirwan JP, Houmard JA, Mitchell JB, Thomas R, Park SH: Effects of repeated days of intensified training on muscle glycogen and swimming performance. Med Sci Sports Exerc 1988;20:249–254.

61 Lamb DR, Rinehard KF, Bartels RL, Sherman WM, Snook JT: Dietary carbohydrate and intensity of interval swimm training. Am J Clin Nutr 1990;52:1058–1063.

62 Sherman MW, Wimer GS: Insufficient dietary carbohydrate during training: does it impair athletic performance? Int J Sports Nutr 1991;1:28–44.

63 Symons JD, Jacobs I: High intensity performance is not impaired by low intramuscular glycogen. Med Sci Sport Exerc 1989;21:550–557.

64 Henriksson J: Training-induced adaptations of skeletal muscle and metabolism during submaximal exercise. J Physiol 1977;270:661–675.

65 Hardman AE, Williams C, Boobis LH: Influence of single-leg training on muscle metabolism and endurance during exercise with trained limb and untrained limb. J Sports Sci 1987;5:105–116.

66 Devlin J, Williams C: Consensus Statement on Foods, Nutrition and Sports Performance. J Sports Sci 1991;9(suppl):iii.

67 Moses K, Manore MM: Developing and testing of carbohydrate monitoring tool for athletes. J Am Diet Assoc 1991;91:962–965.

Prof. Clyde Williams, Department of Physical Education,
Sports Science and Recreation Management, Loughborough University,
Loughborough, Leicestershire, LE11 3TU (UK)

Simopoulos AP, Pavlou KN (eds): Nutrition and Fitness for Athletes.
World Rev Nutr Diet. Basel, Karger, 1993, vol 71, pp 61–68

Contribution of Fats and Fatty Acids to Performance of the Elite Athlete

P.J. Nestel

CSIRO Division of Human Nutrition, Adelaide, S.A., Australia

Introduction

Classic studies of the relative contributions of fat and carbohydrate to total energy consumption during exercise showed clearly that the preceding diet influenced the intensity and duration of physical activity. The R value was lower during exercise following a high fat diet [1]. The time to exhaustion was also less after high fat diets [2]. There was also a progressive shift to oxidizing fat during the course of prolonged exercise, to the point where fatty acids contributed 80–90% of energy, up from an initial 20% [3]. This suggested that the exclusive use of fat could not support the fuel requirements of exercise beyond a relatively meagre 50% of maximal oxygen uptake. When lipid mobilization from fat stores was inhibited with nicotinic acid the capacity for muscle work declined [4].

Whereas the body's store of carbohydrate is meager and needs continuous replenishment, the fat stores are not limiting as a source of energy. If for instance fatty acids were the sole source of energy, less than 1 kg adipose fat would be needed to run the marathon. Since this represents only about 10% of fat stores, there is clearly no need for fat supplementation during training or during exhausting exercise.

Stored lipid contributes energy either directly as fatty acid or indirectly after glycerol is converted to glucose. Fat does not need to be eaten during exercise to provide this substrate nor indeed is this even contemplated. The body's physiology adapts during training and prolonged exercise to enhance the mobilization of lipid for delivery to muscles. This became apparent in the early 1960s soon after free fatty acids (FFA) were identified in circulating blood and shown to have an important energy function, despite their low concentration. However, the turnover of these fatty acids is high, of the order of a few minutes so that substantial amounts can be delivered to muscle. Control occurs through metabolic events in adipose tissue where the breakdown of triglyceride to FFA and glycerol is regulated. A major stimulus for the lipolysis of this triglyceride is the sympathetic nervous system and its circulating catecholamines. In 1964, Steinberg et al. [5] demonstrated the link between the flux of FFA and energy utilization when they infused norepinephrine into subjects to mobilize FFA and observe a corresponding rise in oxygen uptake and CO_2 production. At about the same time, a number of investigators showed that exercise stimulated plasma FFA turnover and an increase in limb muscle uptake of FFA which was proportional to the raised concentration of FFA [6].

Subsequent research demonstrated the mechanisms responsible for exercise-induced FFA mobilization. Stored triglyceride is catabolized by a hormone-sensitive lipase. It is stimulated by at least two of the major hormones which are rapidly secreted when exercise begins, catecholamines and growth hormone. Insulin, the most potent inhibitor of this lipase, is itself suppressed by exercise (probably by the rising level of norepinephrine). The net hormonal effect is to increase lipolysis within adipocytes and stimulate FFA efflux. At the same time, the sensitivity of adipocytes to epinephrine increases [7].

Not all fat stores behave identically. Deep abdominal fat cells are most sensitive to hormone-stimulated lipolysis. This may be relevant to athletes who carry excess fat and to the nature of their adiposity. Fat around the thighs may mobilize less FFA than that in the abdomen. Further, not all adrenoreceptors (which respond to catecholamines) stimulate lipolysis. Some are inhibitory and their distribution in fat cells is not uniform. In fact, the differing responsiveness of gluteo-femoral (thigh) fat and deep visceral (abdominal) fat may lie in the preponderance of one or other population of adrenoreceptors. Therefore, to the extent that fat may be utilized for energy during exercise, its distribution in stores may be an important factor.

Events during Exercise

The role of FFA in supplying energy to working muscle was established by Swedish physiologists in the early 1960s [4, 6]. Within minutes of the commencement of exercise, the plasma FFA concentration declines steeply despite a rapid rise in its turnover. The more intense the exercise the greater the fall in arterial FFA levels as extraction by muscle increases. This reflects events in adipose tissue and in muscle. The rise in catecholamines and fall in insulin levels switch adipose triglyceride metabolism from storage to breakdown so that FFA are released. In muscle, FFA uptake is increased partly due to greater availability and partly due to opening up of the capillary beds. Heart rate and FFA utilization are proportional [4]. Arteriovenous sampling across large working muscles show that the increased production of CO_2 is to a large measure derived from FFA oxidation [8]. That it is indeed the exercising muscle mass which is responsible for the large increase in FFA oxidation becomes apparent at the end of exercise when plasma FFA concentration rises rapidly and steeply.

There are additional important sources of fatty acid for working muscles. Triglyceride is also stored between muscle fibres, and biopsy of muscles at the end of prolonged intense exercise shows marked depletion of these local stores. Circulating triglyceride within very low density lipoproteins (VLDL) is another important source of energy.

Physical training leads to a higher activity of lipoprotein lipase, the enzyme responsible for VLDL triglyceride catabolism to FFA [9]. Other adaptations to training include heightened beta-oxidation of FFA in muscle. In fact muscles can adapt to utilize lipid to the point where moderately intense exercise (60% of maximum aerobic capacity) can be performed on a very low carbohydrate diet. This leads to better clearance of circulating triglyceride [10]. Phinney et al. [11] studied fat people undergoing a weight-reducing program on a hypocaloric, ketogenic diet. Exercise training over several weeks led to muscles utilizing mainly ketones, estimated from the high circulating ketone levels and the very low respiratory quotient.

After exercise, when glucose utilization drops, especially with prolonged exercise and depletion of muscle glycogen, energy is increasingly supplied by lipid. The concentrations of plasma FFA, lactate and ketonic acids rise significantly for several hours and R falls [12]. This accounts partly for the energy consumed during the excess postexercise oxygen consumption phase.

Effect of Current Dietary Practice

The early demonstrations of the importance of FFA to muscle energy were carried out in subjects eating usual diets and fasting on the days of study. The question is whether lipid remains a substantial energy substrate with current dietary preparations. It is common for athletes to eat a preponderance of carbohydrate during the 3 days before the event, have a large carbohydrate meal 3–4 h beforehand and often take supplementary liquid sugars during a prolonged athletic event. A large amount of carbohydrate is advised following the event to replenish glycogen stores.

These modern dietary concepts lessen but do not abolish reliance on lipid. The higher levels of plasma insulin especially if so-called high glycemic starches and sugars are eaten, will suppress FFA mobilization. Whether the use of low-glycemic foods on the day of the event (in order to provide a slow but lengthy absorption of glucose) will allow greater utilization of FFA, since insulin secretion is less, is not known.

There is, however, no question that lipids are substantial sources of energy in prolonged lengthy exercise: (1) the plasma VLDL concentration falls; (2) the HDL cholesterol concentration rises acutely (reflecting VLDL catabolism); (3) muscle triglyceride becomes depleted; (4) there is loss of body fat.

Effects of Dietary Fat

The energy needs of athletes are high, resulting in a high intake of fat [13]. Modern dietary practice advises a relatively low carbohydrate intake prior to the final 3 days before the event; protein and fat consumption is therefore increased. The value of physical training on the disposal of dietary fat was demonstrated by Cohen et al. [14] who showed that athletes cleared postprandial fat particles more quickly than untrained subjects.

Does the high fat intake have any adverse consequences and does the nature of the fat matter? Plasma cholesterol is influenced but major rises are seen mainly in athletes who practice body building. Different species of eicosanoids are generated from different types of polyunsatured fatty acids, but whether this affects performance has not been reported. Impairment of insulin sensitivity is a possible adverse outcome of a high fat diet.

Optimal glucose utilization and oxidation are goals of training. This occurs partly through improved sensitivity of muscle to insulin. The number and activity of glucose transport proteins also increase with exercise. Tissue sensitivity to insulin affects triglyceride uptake as well as that of glucose. Baselmans et al., [6] showed that factors such as obesity and hypertriglyceridemia which impaired glucose removal during infusions of insulin also retarded triglyceride utilization. Even moderate amounts of training such as a ½ h at 60–70% of Wmax 3 × weekly are enough to improve insulin induced glucose utilization [15] which may therefore be a factor in lowering plasma triglyceride.

High fat diets are known to impair insulin sensitivity and glucose utilization, Swinburn et al. [16] compared the effects of a 50% fat, 30% carbohydrate intake with that of 15% fat, 70% carbohydrate. Oral glucose tolerance was impaired due to reduced glucose-induced removal; insulin-induced removal of glucose was not changed. Sparti and Decombaz [17] showed that a high fat diet reduced glucose oxidation despite prior strenuous exercise, which generally improves glucose utilization. There is therefore good evidence that high lipid availability impairs glucose oxidation in muscle, as initially proposed by Randle et al. [18] nearly three decades ago and confirmed by others [19]. However, it is possible that the inclusion of n–3 fatty acids will negate the adverse effect of a high fat diet [20]. In experimental animals this reflected a lowering of triglyceride in skeletal muscle and enrichment of membranes with n–3 fatty acids [21].

Effects on Plasma Lipoproteins

The major effects of regular exercise are lowering of plasma triglyceride and raising of HDL cholesterol. Plasma total cholesterol is less commonly affected but lowering has been documented. The effects on triglyceride and HDL are linked in that, first, the catabolic products of VLDL generate HDL and this has been demonstrated by Kiens and Lithell [22] across the circulation of the exercising thigh muscles of training athletes. Second, the core lipids exchange between the two lipoproteins leading to an inverse relationship in their lipid concentrations. The increase in skeletal muscle lipoprotein lipase activity through exercise [22] partly brought about by improved insulin sensitivity, is a major cause of the changes in VLDL and HDL concentration. In fact, these changes (rises in lipoprotein lipase and HDL, and a fall in triglyceride) occur following a single marathon run [23].

However, this does not fully explain the high HDL levels which characterize physical fitness. The greater the intensity and duration of exercise, the greater is the rise in HDL [24]. Differences in dietary fat do not explain higher HDL in athletes, although high egg consumption, which raises HDL cholesterol, did correlate significantly with HDL in one study in runners [25]. There is, however, some evidence that in marathon runners, who drink a reasonable amount of alcohol, the high HDL levels are partly due to this habit [26]. Another factor is the usual leanness of athletes [27]; plasma triglyceride are inversely and HDL cholesterol directly related to fat free mass [28]. Weight loss contributes to the rise in HDL [29].

Otherwise, the role of diet appears minimal; athletes tend to consume large amounts of carbohydrate (which generally lowers HDL levels in non-athletes), yet have high HDL concentrations. The increase in HDL concentration mainly reflects HDL_2, a function of increased lipoprotein lipase activity and also of decreased hepatic triglyceride lipase activity which regulates the conversion of HDL_2 to HDL_3.

Endurance training leads not only to loss of body fat but also to changes in adipocyte morphology, the mass of individual fat cells decreasing [7]. However, the fatty acid composition of adipose tissue remains stable, which is a further indication that athletes generally do not change their diets qualitatively to a substantial degree.

Conclusions

High fat diets increase dependence on fatty acid as a major source of fuel during prolonged, intense exercise. This reduces the duration and capacity of performance. High fat diets also worsen glucose utilization. These have been the major reasons for the modern dietary concepts behind athletic performance.

However, fat stores contribute importantly to energy and become depleted with endurance exercise. Replenishment of fat stores is accomplished most readily by eating fat. How much fat and where in the dietary cycle preceding the athletic event are important questions. Whether the type of fat matters may also be important, with respect to glucose utilization and possibly because of the species of eicosanoids generated.

The effects on the HDL concentration are reviewed; dietary factors are probably minor other than in relation to body composition and possibly also to the large amount of alcohol consumed by many athletes.

References

1   Krogh A, Lindhard J: The relative value of fat and carbohydrate as sources of muscle energy. Biochem J 1920;14:290–363.
2   Bergstrom J, Hermansen L, Hultman E, et al: Diet, muscle glycogen and physical performance. Acta Physiol Scand 1967;71:140–150.

3 Edwards HT, Margaria R, Dill DB: Metabolic rate, blood sugar and the utilization of carbohydrate. Am J Physiol 1934;108:203–209.

4 Pernow B, Saltin B: Availability of substrates and capacity for prolonged exercise in man. J Appl Physiol 1971;31:416–422.

5 Steinberg D, Nestle PJ, Elsworth R, et al: Calorigenic effect of norepinephrine correlated with plasma free fatty acid turnover and oxidation. J Clin Invest 1964;43: 167–176.

6 Baselmans J, Nestel PJ, Nolan C: Insulin-induced glucose utilization influences triglyceride metabolism. Clin Sci 1983;64:511–516.

7 Despres JP, Bouchard C, Savard R, et al: The effect of a 20-week endurance training program on adipose-tissue morphology and lipolysis in men and women. Metabolism 1984;33:235–239.

8 Havel RJ, Pernow B, Jones NL: Uptake and release of free fatty acids and other metabolites in the legs of exercising men. J Appl Physiol 1967;23:90–95.

9 Nikkita EA, Taskinen M-R, Rehunen S, et al: Lipoprotein lipase activity in adipose tissue and skeletal muscle of runners: relation to serum lipoproteins. Metabolism 1978;27:1661–1671.

10 Sasy SP, Cullinane EM, Saritelli A, et al: Elevated high density lipoprotein cholesterol in endurance athletes is related to enhanced plasma triglyceride clearance. Metabolism 1988;37:568–572.

11 Phinney SD, Horton ES, Sims EA, et al: Capacity for moderate exercise in obese subjects after adaptation to a hypocaloric, ketogenic diet. J Clin Invest 1980;66: 1152–1161.

12 Withers RT, Gore CJ, MacKay MH, et al: Some aspects of metabolism following a 35km road run. Eur J Appl Physiol Occup Physiol 1992; in press.

13 Poehlman ET, McAuliffe TL, Van Houten DR, et al: Influence of age and endurance training on metabolic rate and hormones in healthy men. Am J Physiol 1990;259: E66–E72.

14 Cohen JC, Noakes TD, Benarde AJS: Postprandial lipemia and chylomicron clearance in athletes and in sedentary men. Am J Clin Nutr 1989;49:443–447.

15 Jennings G, Nelson L, Nestel PJ, et al: The effect of changes in physical activity on major cardiovascular risk factors, hemodynamics, sympathetic function, and glucose utilization in man: a controlled study of four levels of activity. Circulation 1986;73:30–40.

16 Swinburn BA, Boyce VL, Bergman RN, et al: Deterioration in carbohydrate metabolism and lipoprotein changes induced by modern, high fat diet in pima Indians and Caucasians. J Clin Endocrinol Metab 1991;73:156–165.

17 Sparti A, Decombaz J: Effect of diet on glucose tolerance 36 hours after glycogen-depleting exercise. Eur J Clin Nutr 1991; in press.

18 Randle PJ, Garland PB, Hales CN, et al: The glucose fatty acid cycle: Its role in insulin sensitivity and the metabolic disturbances of diabetes mellitus. Lancet 1963; i:785–789.

19 Nestel PJ, Carroll KF, Silverstein MS: Influence of free-fatty-acid metabolism on glucose tolerance. Lancet 1964;ii:115–117.

20 Storlien LH, Kraegen EW, Chisholm DJ, et al: Fish oil prevents insulin resistance induced by high fat feeding. Science 1987;237:885–888.

21 Carruthers A, Melchior DL: Effects of lipid environment on membrane transport: the human erythrocyte sugar transport protein/lipid bilayer system. Ann Rev Physiol 1988;50:257–271.

22 Kiens B, Lithell H: Lipoprotein metabolism influenced by training.induced changes in human skeletal muscle. J Clin Invest 1989;83:558–564.

23 Kuusi T, Kostiainen E, Vartiainen E, et al: Acute effects of marathon running on levels of serum lipoproteins and androgenic hormones in healthy males. Metabolism 1984;33:527–531.

24 Lehtonen A, Viikari J: Serum triglycerides and cholesterol and serum high density lipoprotein cholesterol in highly physically active men. Acta Med Scand 1978;204: 111–114.

25 Hartung GH, Foreyt JP, Mitchell RE, et al: Relation of diet to high density lipoprotein cholesterol in middle aged marathon runners, joggers, and inactive men. N Engl J Med 1980;302:357–361.

26 Willet W, Hennekens CH, Siegel AJ, et al: Alcohol consumption and high density lipoprotein cholesterol in marathon runners. N Engl J Med 1980;303:1159–1161.

27 Moore CE, Hartung GH, Mitchell RE, et al: The relationship of exercise and diet on high density lipoprotein cholesterol levels in women. Metabolism 1983;32:189–196.

28 Wood PD, Haskell WL, Blair SN, et al: Increased exercise level and plasma lipoprotein concentrations: A one-year, randomized, controlled study in sedentary, middle-aged men. Metabolism 1983;32:31–39.

29 Williams PT, Wood PD, Krauss RM, et al: Does weight loss cause the exercise-induced increase in plasma high density lipoproteins? Atherosclerosis 1983;47:173–185.

P.J. Nestel, MD, FRACP, FTS, CSIRO Division of Human Nutrition,
PO Box 10041, Gouger Street, Adelaide, S.A. 5000 (Australia)

Concerns for Specific Population Groups in Relation to Nutrition and Fitness

Simopoulos AP, Pavlou KN (eds): Nutrition and Fitness for Athletes.
World Rev Nutr Diet. Basel, Karger, 1993, vol 71, pp 69–83

The Body Profile Analysis System (BPAS) to Estimate Ideal Body Size and Shape: Application to Ballet Dancers and Gymnasts

Frank I. Katch

Department of Exercise Science, University of Massachusetts,
Amherst, Mass., USA

Introduction

Dancers and gymnasts are often preoccupied with maintaining an ideal body size and shape for purposes of 'looking good' during high level performance and competition. Such a priority is understandable because looking too heavy (above ideal weight) or too 'fat' (protruding stomach, large thighs, hips, buttocks) would simply not be tolerated, not even in junior level competition involving prepubescent participants. An athlete who deviates from a preconceived ideal for body size would not be allowed by their coach or trainer to continue with the sport, let alone perform in public. Thus, prospective athletes and seasoned performers are under immense pressure to achieve an ideal body physique. To accomplish such ends, the athlete voluntarily restricts nutrient intake (dieting) to curtail body weight gain and accumulation of excess body fat. This is often associated with lower than normal energy intake [1–3], nutrient deficiencies, especially in minerals [3–6], endocrine disturbance related to menstrual irregularities [7–9], the common eating disorders anorexia nervosa and bulimia [10, 11], and increased risk for skeletal and bone injuries [12–15].

Background

Many approaches can be used to assess physique status for purposes of body composition evaluation. Whatever technique is used, a fundamental issue confronting both the coach and athletes is the intervention strategy to alter body composition. For dancers and gymnastics, this means reducing body fat (and most likely body weight). The question then becomes by how much, and therein lies the dilemma. At present, there is no way to know what is most desirable in terms of body fat content as it relates to performance. For example, if body mass is lower by 3 kg, or body fat is less by 3%, will these differences really have a significant impact on final performance? There are probably as many opinions on the matter as there are coaches and trainers. What is known for certain is that coaches and athletes will attempt to do whatever they believe it takes to win, and this includes dietary manipulation to alter body composition.

It is inconceivable to think that coaches and sport scientists would counsel their athletes that winning or achieving top performance is unimportant. On the contrary, striving for lofty goals is worthy and noble without need for social or moral justification. What seems unfortunate, however, is the perception that successful performance at all levels of competition requires an ideal body shape with the lowest possible body fat content. One corollary of such a perception is that nutritional restraint provides the predominant means to unbalance the energy balance equation to achieve an ideal physique. but who decides what constitutes the ideal physique?

One obvious answer is the top performer who wins the competition or makes the select team! While the athlete may not verbally advertise that 'thinness' and low fat content are desirable, their appearance during competition in fact does 'advertise' the importance of physique status. The after effect is immediate and long lasting because successful performance translates to media advertising. The svelte and graceful gymnast or dancer who is displayed in magazine and newspaper ads, or who endorses commercial products on TV or video, does in fact, subtly 'sell' thinness to whatever audience is targeted for the advertising. This is also true of the performance per se. Except for a relatively few top performers who are able to compete at the international level, the thousands of other participants in ballet and gymnastics strive to become like the champions, not only in skill but in physique and body composition.

In previous anthropometric studies of the body composition of ballet dancers and gymnasts [6, 16], the approach has been to use athletes in the respective sports as the frame of comparison. For example, when evaluating adolescent female gymnasts, their physique status is often compared to other adolescent women gymnasts [6, 16]. While this may be helpful in some sports such as American football or track, where excess body fat may hinder performance, this is usually not the case in ballet and gymnastics where thinness and avoidance of body fat is predominantly the goal.

The purpose of the present paper is to describe a new application of an anthropometric technique to evaluate body size and shape called the Body Profile Analysis System (BPAS). The BPAS does not require conformity to standards established for dancers and gymnasts; thus, the BPAS diminishes the need to perpetuate the struggle to achieve an ideal that rewards underweightness and directly and indirectly contributes to medical problems. Instead, criteria based on the normal population according to age, gender, and race would lessen the need to emulate the physique of the elite performer.

Development of the Body Profile Analysis System

The Somatogram

Behnke et al. [17] in 1959 were the first to describe the relative size of the trunk and extremities expressed in percentage deviation units from reference standards developed from large-scale anthropometric surveys of the military and civilian populations. The original body profile was referred to as the somatogram (Som), and the basis of Som was the body size module ($F = \sqrt{\text{Body weight, kg}}$/stature, dm). This quantity allowed a conversion between a squared matrix of 12 girths and the square root of the quantity body mass (kg) divided by stature (dm). For an individual or group, the 12 girths for Som were displayed graphically as percentage deviations from the reference values for the 12 girths. For example, to construct the Som for an individual, each of 12 girths (g) are divided by their proportionality constants (k) derived from a reference group [18], to obtain a deviation (d) score, where $d = g/k$. The k constant is computed as g/D, where D equals the sum of the g values divided by 100. The reference values are gender specific, and form the basis of the Behnke Reference Man and Reference Woman [19].

The graphic representation of Som is the percentage deviation of each d quotient from D. This graphic approach has been used to show changes in body size during the growth period, to show how body shape changes with aging, and to describe gender differences among athletic groups [20].

Measurement of Girths

A cloth measuring tape is applied lightly to the skin surface so the tape remains taut but not tight. Duplicate measurements are taken at each site and the average is used as the criterion circumference score. A metal tape is not recommended because of skin compression. Reliability of measurement for all sites should exceed $r = 0.98$. The anatomic landmarks for the 12 girth sites are taken as follows:

1 Shoulders: maximal protrusion of the bideltoid muscles and the prominence of the sternum at the junction of the second rib.
2 Chest: for men about one inch above the nipple line; for women at the axillary level. Note: in men and women, the tape is placed in position with the arms held horizontally. The arms are then lowered and the measurement recorded at the mid-tidal level of respiration.
3 Abdomen 1: the conventional circumference of the waist just below the rib cage at the minimal width.
4 Abdomen 2: level of the iliac crests at the navel. Note: the average of abdomen 1 and abdomen 2 is used as the abdominal measurement.
5 Buttocks: maximal protrusion and, anteriorly, the symphysis pubis. The heels are kept together.
6 Thighs: crotch level at the gluteal fold.
7 Biceps: maximal circumference with the arm fully flexed and fist clenched.
8 Forearms: maximal circumference with the arm extended and palm up.
9 Wrists: the circumference distal to the styloid processes of the radius and ulna.
10 Knees: the middle of the patella with the knee relaxed in slight flexion.
11 Calves: maximal circumference.
12 Ankles: minimal circumference just above the malleoli.

Note that in the calculation of Som, the abdomen 1 and abdomen 2 measurements are averaged and plotted as abdomen average.

Original Somatogram Calculations

The left side of table 1 lists the girth measurements and k constants for the Reference Man and Women Som originally developed by Behnke et al. [17]. To calculate Som, each girth (g) is divided by k to obtain the ratio d ($d = g/k$). The reference value is then computed as D, where D is the sum of the girths (D = sum girths) divided by the sum of the k values (sum k = 100). The graphic representation of body shape is plotted as percentage deviations of each d from D [% deviation = $(d - D)/D$]. If an individual's

Table 1. Proportionality constants (k) for the Som for the Reference Man and Woman, and body profile analysis k values using PE-M and PE-NM components

	Reference Man (Som)		Reference Woman (Som)		Reference Man PE-M, k	Reference Woman PE-M, k
	girth, cm	k	girth, cm	k		
Muscular girths						
Shoulders	110.8	18.47	97.4	17.51	55.40	52.59
Chest	91.8	15.30	82.5	14.85	45.90	44.55
Biceps	31.7	5.29	26.7	4.80	15.85	14.42
Forearm	26.9	4.47	23.1	4.15	13.45	12.47
Thigh	54.8	9.13	55.8	10.03	27.4	30.13
Calf	35.8	5.97	34.1	6.13	17.90	18.41
Sum	351.8	58.63	319.6	57.47	175.9	172.57
					PE-NM, k	PE-NM, k
Non-muscular girths						
Abdomen 1	77.0	12.84	65.6	11.83	38.50	35.42
Abdomen 2	79.8	13.30	77.8	13.95	39.90	42.00
Abdomen, average	78.4	13.07	71.7	12.90	39.20	38.71
Hips	93.4	15.57	94.2	16.93	46.70	50.86
Knees	36.6	6.10	34.9	6.27	18.30	18.84
Wrists	17.3	2.88	15.2	2.73	8.65	8.21
Ankles	22.5	3.75	20.6	3.70	11.25	11.12
Sum	248.2	41.37	236.6	42.53	124.1	127.74
Total sum	600.0	100.0	556.2	100.0	300.0	300.3

measurements conform precisely to the Reference values, there is no deviation and the Som would plot as a vertical line. For example, for a biceps of 40.2 cm and $D = 6.771$ (sum 11 g/100), d for the biceps would be 7.60 (d = 40.2/5.29), where 5.29 is the k(biceps) value for the Reference Man listed in table 1. Expressed as a deviation from D, d(biceps) would be 12.2% larger (7.600–6.771/6.771) $\times$ 100, and would be plotted on Som as +12.2 to the right of the zero axis. The d values for the remaining girth sites are then plotted in a similar fashion. Figure 1 shows an example of the original Som for a collegiate female gymnast following 16 months of bed rest due to a sports injury. In this example, the shoulder, chest, abdomen average, thigh, biceps, and forearm have positive % deviations (with the

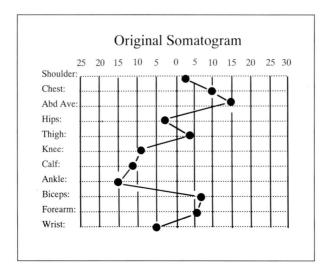

Fig. 1. Original somatogram. Data are from a collegiate female gymnast following 16 months of bed rest due to a sports injury. The athlete had gained 17 kg body mass, and her body fat percentage assessed by hydrostatic weighing increased from 14.2 to 22.9%.

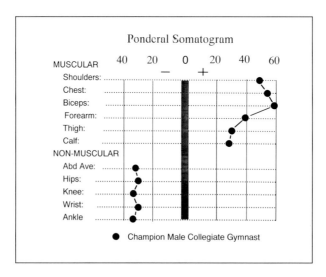

Fig. 2. Ponderal somatogram for a world champion male collegiate gymnast. Note the excessive muscular development in the biceps, chest, and shoulders, and the relatively 'smaller' nonmuscular regions. The ratio PE-M/PE-NM = 1.522.

abdomen the largest at 15.0%), while there are negative % deviations for the hips, knee, calf, ankle, and wrist (the ankle has the lowest % deviation of 15%) Presumably, the upper trunk and arms were the sites of excessive fat accumulation during convalescence.

One criticism of the Som approach was that it did not permit translation of girth size into a volume or weight entity that related to the body as a whole. The original Som also did not differentiate between muscular and nonmuscular areas of the body; thus, nonmuscular girths such as the abdomen and hips were integrated with muscular parts such as the flexed biceps, thigh, and calf. Because the deviation of each d from D is based on the matrix of girths, each g is somewhat related to itself because it is part of D. While this is probably of minor importance when graphing the d values for Som, it still does not permit a clear-cut separation of the muscular and nonmuscular components.

The Body Profile Analysis System

The BPAS is an extension of the Som. Girths are converted to ponderal equivalent weight values and presented as a ponderal Som (PSom) The matrix of girths are separated into muscular and nonmuscular components and compared as mass equivalents. An example of PSom is presented in figure 2 for a world champion male collegiate gymnast where there is excessive muscular development, especially in the biceps, chest, and shoulders.

PSom Calculation

The right side of table 1 lists the k constants for the Reference Man and Reference Women that are used to calculate the PSom. There are two components:

(1) Ponderal equivalent muscular component (PE-M) that includes the shoulder, chest, biceps, forearm, thigh and calf.

(2) Ponderal equivalent nonmuscular component (PE-NM) that includes two abdominal measures and their average, hips, knee, wrist, and ankle.

The constants for the individual girths are calculated from the data of the Reference Man and Woman as $k = g/F$, where g = individual girth in cm, and F = the square root of the Reference Man and Women median weight in kg divided by Reference Man and Woman median stature in dm.

The value of F for the Reference Man is 2.000 and 1.852 for the Reference Woman [17].

The PE for each girth, expressed in kg, is computed as the square of the quotient g/k multiplied by stature in dm. For example, the PE for the shoulders for the Reference Man is $(g/k)^2 \times$ stature or $(110.8/55.4)^2 \times 17.4 = 69.6$ kg. For the Reference Man, that PE values for all of the girths are identical to the reference median weight of 69.6 kg; the same is true for the Reference Woman. All of the PE values for the girths are identical to the Reference Woman median weight of 56.2 kg. For the reference models, the deviations of each PE from their respective standards are necessarily zero because there is no deviation from group symmetry. In using the BPAS, the reference values based on population data represent the standard.

Step-by-Step Procedure to Calculate the Body Profile

The step-by-step procedure to calculate the body profile uses the k constants in table 1. These data are then applied to the data for a male gymnast in table 2.

– Step 1: Compute PE for the muscular girths using the constants displayed at the left side of tables 1 and 2. Compute PE for each girth as follows:

PE girths = $(g/k)^2 \times$ Ht, dm, where g is the girth (cm) and k is the reference constant for g.

For example, the PE for the shoulders in table 2 is computed as $(128.8/55.4)^2 \times 17.06 = 92.2$ kg, where shoulder k = 55.4 (Reference Man), 128.8 cm is shoulder girth, and 17.06 is stature, dm. Repeat step 1 for the other M girths in table 2 using the respective g and k values.

– Step 2: Compute the average PE for the M girths. In table 2, the average PE for the M girths is 100.4 kg.

– Step 3: Compute the sum of the M girths. In table 2, the sum of the M girths is 421.9 cm.

– Step 4: Compute the average PE for the NM girths. For this example, the average PE-NM = 71.9 kg.

– Step 5: Compute the sum of the NM girths. For this example, the sum of PE-NM = 253.6 cm.

– Step 6: Compute the following:

(Step 6/sum NM g for Ref standard)$^2 \times$ Ht, dm.

For this example, $(253.6/124.1)^2 \times 17.06 = 71.2$ kg, where 124.1 is the sum of the NM g for the Reference Man.

Table 2. Calculation of PSom for a male gymnast whose body mass = 84.2 kg, stature = 17.06 dm, and F = 2,222, where F = √body mass, kg/stature, dm

Reference Man k	Site	Girth, cm	g/k	PE-M	% dev from 71.9 kg
55.4	shoulder	128.8	2.325	92.2	28.2
45.9	chest	115.1	2.508	107.3	49.2
15.85	biceps	43.4	2.738	127.9	77.9
13.45	forearm	33.1	2.461	103.3	43.7
27.40	thigh	61.1	2.230	84.8	17.9
17.90	calf	40.4	2.257	86.9	20.9
175.9	Sum	421.9		X = 100.4	
				PE-NM	% dev from 100.4 kg
39.20	abdomen, average	80.3	2.048	71.6	28.7
46.70	hips	94.7	2.028	70.2	30.1
8.65	wrists	18.0	2.081	73.9	26.4
18.30	knees	37.4	2.044	71.3	29.0
11.25	ankles	23.2	2.062	72.5	27.7
124.1	Sum	253.6		X = 71.9	

Note that the % dev for PE-M is computed by using the average of the PE-Nm (71.9; step 7 in the example; the % dev for PE-NM is computed by using the average of the PE-M (100.4; step 8 in the example). To compute PE-M or PE-NM, first divide the girth by its k constant. The quotient is squared and multiplied by stature, dm. For the shoulder, for example, 128.8 / 55.4 = 2.325. PE-M is then $(2.325)^2 \times 17.06 = 92.2$ kg. The % dev for the PE-M for the shoulder is +28.2. This is computed as $[(92.2-71.9)/71.9] \times 100$.

– Step 7: Compute % dev for each of the M girths as:
   $(PE\text{-}M_{(girth)}$ minus PE-NM)/PE-NM.
   For the shoulder, % dev is calculated as follows:
   % dev$_{(shoulder)}$ = (92.2–71.9)/71.9 = +28.2.
– Step 8: Compute % dev for each NM girth as:
   $(PE\text{-}NM_{(girth)}$ minus PE-M)/PE-M $\times$ 100.
   For the hips, % dev is calculated as follows:
   % dev$_{(hips)}$ = (70.2–100.4)/100.4 $\times$ 100 = –30.1.

–   Step 9: Plot the % dev for each of the remaining girths in steps 7 and 8.

If one is interested in computing an overall total deviation (Tot dev), without regard to M and NM, then the sum of the M and NM girths are used in the following equation:

Total dev = (total G, cm/total k)^2 × Ht, cm.

For the data in table 2, Tot dev is computed as $(675.5 \text{ cm}/300)^2$ × 17.06 = 86.5 kg, where 675.5 cm is the sum of 11 girths for the gymnast, and 300 is the sum of the k values for PE-M and PE-NM. To compute an average for PE-M or PE-NM, use the following equation:

(Step 3/sum g for Ref standard)^2 × Ht, dm.

In table 2, for example, the average PE-M is computed as $(421.9/175.9)^2$ × 17.06 = 98.1 kg, where 175.9 is the sum of the muscular g for the Reference Man, and 421.9 is the sum of the muscular girths for the gymnast.

A unique aspect of the new BPAS is the calculation of the d values. In the original Som, there was no separation of the girth matrix into muscular and nonmuscular components because Som was calculated as the sum of the girths/100. Thus, a specific d value was related to the sum of the girths that included that particular girth.

This complication has been avoided in BPAS by comparing the PE-M girth values with the average of the PE-NM values, and vice versa. There will not be exact numerical equivalency between the total (cm/k)^2 × stature and the average of the PE values because of differences in proportionality between the Reference Man and Woman, and among individuals or groups of individuals. For the BPAS displayed in figure 2, the specific k values were from the values of BPAS listed for the Reference Man in table 1. Note that the % deviations for the gymnast in figure 2 differ only slightly from the % dev for the gymnast whose measurements are displayed in table 2. The largest difference is in the biceps.

Comments on Specific Applications of BPAS for Gymnasts and Dancers

While there have been numerous studies of the gross body composition and physique attributes of outstanding male and female gymnasts and ballet dancers [21–24], there is only limited anthropometric data that can be used to compute Som and PSom, especially at the championship and

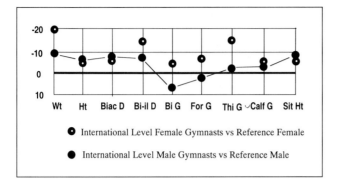

Fig. 3. Selected physique dimensions of international level male and female gymnasts with the corresponding dimensions of the Reference Man and Woman. Percent differences. Wt = Weight, kg; Ht = height, cm; Biac-D = biacromial diamter, cm; Bi-ilD = bi-iliac diameter, cm; BiG = bicepts girth, cm; ForG = forearm girth, cm; ThiG = thigh girth, cm; CalfG = calf girth, cm; and SitHt = sitting height.

international levels. In one of the few studies where the full complement of girths was taken [16], the argument was made that the measurements for a reference ballet dancer should be used instead of comparisons with Behnke's Reference Woman. Freedson [16] states: 'It is of utmost importance that the so-called ideal physique for the classical ballet dancer not only should consider maximizing thinness to complete the necessary visual aesthetic qualities but must also be related to maintenance of health and well being.' There is no question that as long as the criterion of body composition is that of the competitive athlete, the coach *and* athlete will strive to achieve an ideal that exemplifies the characteristics of the champion performer. This is given further credence by the researcher who uses the champion as the role model for physique status. The opposing point of view would allow the reference standard to be the Reference Woman. Perhaps this would reduce the quest of aspiring athletes to try and attain the physique dimensions of the top-level performer.

Figure 3 compares selected physique dimensions of international level male and female gymnasts with the corresponding dimensions of the Reference Man and Woman. The international male and female competitors are 'smaller' than the reference standards for each of the body dimensions, except for biceps and forearm girth, which are larger for the male gymnasts compared to the Reference Man.

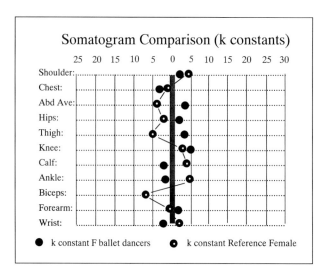

Fig. 4. Somatogram comparison for female ballet dancers using k constants from a reference group of ballet dancers and the Reference Woman. F = Female.

Figure 4 illustrates the point that using the k values based on a group of athletes (in this case female ballet dancers), changes Som dramatically in terms of a 'snapshot' of physique. In this particular example based on the data of Clarkson et al. [6] for adolescent female ballet dancers, using k values for ballet dancers makes some girth sites 'larger' and may encourage bizarre eating behaviors to try and alter physique status. Note that the abdominal and hips regions, two problem areas of concern, actually become positive deviations, while the calf site becomes negative. In contrast, using the k constants for the Reference Woman results in negative deviations for the abdomen, hips, and thigh regions. From a purely psychological perspective, one may tend to down play the importance of further reductions in these areas if they are not targeted as 'too large' based on a particular group of athletes measurements. Note that regardless of which set of k values are used, the actual girths would remain unchanged. Thus, it is only the 'snapshot' of the matrix of girth proportions that has the impact on whether the athlete should attempt to change the dimensions of a particular girth area.

While the Som is a generally useful tool for anthropometric assessment, the new BPAS permits separation of the 11 girths into muscular and nonmuscular components. Furthermore, the BPAS allows the girth data to be interpreted in terms of mass (ponderal) equivalents. For the girth measurements of the male gymnast presented in table 2, the excessive muscular development becomes evident for the six muscular sites, in particular the ponderal equivalent for the biceps of 127.9 kg. In percent deviation terms, the % dev of 77.9% for the biceps is the largest of the six muscular sites. The absolute value of 127.9 kg is 43.7 kg larger than the athletes body mass of 84.2 kg, and 56.0 kg larger than the average PE-NM. This can be interpreted to mean that the mass of the biceps at a body mass of 84.2 kg is the projected size of the biceps if the gymnast weighed 127.9 kg. The other values can be interpreted in similar fashion. Thus, for example, the PE-NM for the hips of 70.2 kg means that this gymnast has a hip girth that should correspond to a body mass of 70.2 kg. The discrepancy of 14.0 kg between the ponderal equivalent for the hips and body mass indicates that the hip region is indeed 'undersized' in relation to the proportions of the other 'oversized' muscular regions, but close to the average PE-NM of 71.9.

To the extent that the PE-M is greater than PE-NM provides a crude estimate of excess muscle. For the BPAS of the well developed male collegiate gymnast displayed in figure 2, the ratio is 1.522 between PE-M and PE-NM. In contrast, PE-M/PE-NM is slightly positive (ratio = 1.023 for the female ballet dancers whose girth deviations are shown for Som in figure 4. That is, the sum of the ponderal equivalents for the muscular girths exceeded the sum of the ponderal equivalents for the nonmuscular girths. For the professional ballet dancers reported in Behnke [18], the ratio of PE-M/PE-NM was 1.030.

Conclusion

In summary, the original Behnke Som to quantify body shape has been a useful approach for partitioning a matrix of girths into ponderal equivalent muscular and nonmuscular components that can be related to the body as a whole [20]. In male gymnasts, for example, excess muscular development predominates in the biceps without compensatory hypertrophy in the lower limbs. Even at the extremes that include the massively obese as well as diminutive and large adolescents, there appears to be a fundamental, intrinsic association between an individual's body mass and the squared matrix of girths multiplied by stature. The BPAS should provide an opportunity to evaluate the relative degree of muscularity in a wide range of individuals, not only athletic populations.

A unique aspect of BPAS is that it does not require conformity to standards established for dancers, gymnasts, and other sport groups. The development of specific k constants may be useful as a frame of comparison among athletic groups such as football, soccer, basketball, and swimming; in contrast, the use of the BPAS for gymnasts and dancers based on the Reference Man and Woman would diminish the need to strive to attain the physique and body shape possessed by the elite performer. The perpetual struggle to emulate an ideal that rewards thinness may be the root cause of both acute and chronic medical problems.

References

1   Benson JE, Eiserman PA, Wardlaw GM: Relationship between nutrient intake, body mass index, menstrual function, and ballet injury. J Am Diet Assoc 1989;89:58–63.
2   Benardot D, Schwarz M, Heller DW: Nutrient intake in young, highly competitive gymnasts. J Am Diet Assoc 1989;89:401–403.
3   Calbrese L, et al: Menstrual abnormalities, nutritional patterns, and body composition in female classical ballet dancers. Phys Sports Med 1983;11:86–98.
4   Calabrese LH: Nutritional and medical aspects of gymnastics. Clin Sports Med 1985;4:23–30.
5   Reggiani ER, Arras GB, Trabacca S, et al: Nutritional status and body composition of adolescent female gymnasts. J Sports Med Phys Fitness 1989;29:285–288.
6   Clarkson PM, Freedson PF, Keller B, et al: Maximal oxygen uptake, nutritional patterns and body composition of adolescent female ballet dancers. Res Q Exerc Sport 1985;6:180–184.
7   Garner DM, Garfinkel PE, Rockert W, et al: A prospective study of eating disturbances in the ballet. Psychother Psychosom 1987;48:170–175.
8   Abraham SF, Beumont PJV, Fraser IS, et al: Body weight, exercise and menstrual status among ballet dancers in training. Br J Obstet Gynecol 1982;89:507–510.
9   Frisch RE, Wyshak G, Vincent L: Delayed menarche and amenorrhea in ballet dancers. N Engl J Med 1980;303:17–19.
10 Druss RG: Body image and perfection of ballerinas: Comparison and contrast with anorexia nervosa. Gen Hsop Psychol 1979;2:115–118.
11 Maloney MJ: Anorexia nervosa and bulimia in dancers. Accurate diagnosis and treatment planning. Clin Sports Med 1983;2:549–555.
12 Bejjani FJ: Occupational biomechanics of athletes and dancers: a comparative approach. Clin Podiatr Med Surg 1987;4:671–710.
13 Caine D, Cochrane B, Caine C, et al: An epidemiologic investigation of injuries affecting young competitive female gymnasts. Am J Sports Med 1989;17:811–820.

14 Lindner KJ, Caine DJ: Injury patterns of female competitive club gymnasts. Can J Sport Sci 1989;15:254–261.

15 Goldstein JD: Spine injuries in gymnasts and swimmers. An epidemiologic investigation. Am J Sports Med 1991;19:198–201.

16 Freedson PF: Body composition characteristics of female ballet dancers; in Clarkson PM, Skrinar M (eds): Science of Dance Training. Champaign, Human Kinetics Books, pp 109–124.

17 Behnke AR, Guttentag OE, Brodsky C: Quantification of body weight and configuration from anthropometric measurements. Hum Biol 1959;31:213–234.

18 Behnke AR: Anthropometric evaluation of body composition throughout life. Ann NY Acad Sci 1963;110:75–78.

19 Behnke AR, Wilmore JH: Evaluation and Regulation of Body Build and Composition. Englewood Cliffs, Prentice Hall, 1974, pp 45–84.

20 Katch FI, Behnke AR, Katch VL: The ponderal somatogram: evaluation of body size and shape from anthropometric girths and stature. Hum Biol 1987;59:439–458.

21 Claessens AL, Veer FM, Stijnen V, et al: Anthropometric characteristics of outstanding male and female gymnasts. J Sports Sci 1991;9:53–74.

22 Dolgener FA, Spasoff TC, St John WE: Body build and body composition of high ability female dancers. Res Q Exerc Sport 1980;51:599–607.

23 Faria IE, Faria EW: Relationship of the anthropometric and physical characteristics of male junior gymnasts to performance. J Sports Med Phys Fitness 1989;29:369–378.

24 Benardot D, Czerwinski C: Selected body composition and growth measures of junior elite gymnasts. J Am Diet Assoc 1991;91:29–33.

Dr. Frank I. Katch, Department of Exercise Science, University of Massachusetts, Amherst, MA 01003 (USA)

Simopoulos AP, Pavlou KN (eds): Nutrition and Fitness for Athletes.
World Rev Nutr Diet. Basel, Karger, 1993, vol 71, pp 84–96

Nutritional and Fitness Considerations for Competitive Wrestlers

Charles M. Tipton[a], *Robert A. Oppliger*[b, 1]

[a] University of Arizona, Tucson, Ariz.;
[b] University of Iowa, Iowa City, Iowa, USA

Introduction

Historians and archaeologists have documented from the Beni Hassan excavations in Egypt that wrestling has been known to be a combative or a competitive activity since 3400 BC [1, 2]. To many, the most famous match of all time was between Jacob and the Angel, as noted in Genesis 32 [3]. Wrestling advocates must realize that most of the holds used today were known to ancient wrestlers [2] and, in all probability, the best-known ancient wrestler was Milos of Croton who wore the wild olive crown and palm of a champion on six different occasions [1].

The ultimate in wrestling is to be a participant in the Olympics, and in 1992 there were ten separate weight classes in either Greco-Roman or in freestyle wrestling ranging from 48 to 130 kg (table 1). Unlike the 'early days' when the match would last an hour, the Olympic match now consists of a single period of 5 min. However, we must realize that not all amateur wrestling is confined to international competition as there are national organizations that sponsor and regulate wrestling matches. Regardless of the organizational and competitive level, matches are driven by the intervals between weight classes, the time allowance for the weigh in, the duration of the match, and the number of matches within a given day (table 1).

[1] It is a pleasure to acknowledge the assistance of Ms. Kathleen Wilkin in the preparation and proofing of the manuscript.

Table 1. Profile of amateur wrestling competition

Parameter	Scholastic	Collegiate	International
Weight classes kg	46.8	53.6	48.0
	50.9	57.3	52.0
	54.1	61.4	57.0
	56.8	64.5	62.0
	59.1	68.2	68.0
	61.4	71.8	74.0
	63.6	75.9	82.0
	65.9	80.4	90.0
	69.1	86.4	100.0
	72.7	125.0	130.0
	77.7		
	85.9		
	125.0		
Match rules	three 2-min periods	one 3-min period two 2-min periods	one 5-min period
Matches per day rules	five w/minimum 45-min period between matches	no official rule	no official rule; no more than three are recommended
Weigh-in rules	1 h maximum; half-hour minimum before dual meet	5 h maximum; half-hour minimum unless otherwise agreed	night before competition 2nd weigh-in has 1.5-hour limit starting immediately after last bout of the day
Overtime condition rules	60-second period	one 2-min period; 30-second sudden death period	sudden death until a winner

Values rounded to the next significant figure; other information taken from 1991–1992 National Federation High School Wrestling Rules provided by Mr. Don Herrmann and from data provided by Dr. A.W. Taylor.

These factors, which impact on the training principles and nutritional practices followed by wrestlers, will provide the foundation for this presentation. Although the emphasis will be on North American models, the information and principles being discussed will be appropriate for other countries as well.

Table 2. Percent of energy source during maximum effort for various times

Energy component	Duration of maximum effort						
	0.5 min	1.0 min	2.0 min	3.0 min	4.0 min	5.0 min	7.0 min
Aerobic	20	30	50	60	65	75	80
Anaerobic	80	70	50	40	35	25	20

Values are approximate and have been modified from Astrand and Rodahl [10] and from information published in the first edition of the text.

Fitness Considerations

In accordance with the concept of exercise specificity [4, 5], the training practices necessary to become a successful wrestler are unique to the sport. This means that individuals must be conditioned in the movement patterns of the sport within the time period allowed, and coaches must analyze wrestling for its anatomical, physiological, and metabolic components and then design the practice 'workouts' accordingly. Careful analysis of skilled wrestlers indicates that aspects associated with muscular power-force-endurance, balance, coordination, reflex and movement time, flexibility, and agility must be integrated and mastered in order for success to occur. From the impressive research findings of Taylor, Sharrat and co-workers with wrestlers [2, 3, 6–9], it is evident that both aerobic and anaerobic capacity [10] are important for participants in amateur wrestling events because a single period can last from 30 s to 5 min, depending on the circumstances (tables 1, 2).

From a theoretical perspective, an individual with a predominance of fast-twitch muscle fibers should be more successful than one with only slow-twitch fibers, although a mixture of the two types is the practical solution. In fact, this seems to be the situation with successful international wrestlers [6–8]. Because the use of muscle-biopsy procedures should be used only by qualified medical scientists on elite athletes in order to study significant physiological and biochemical problems, the overwhelming majority of wrestling coaches and their medical advisors must use conventional methods and procedures to select members of their teams (speed, power, endurance).

It is our belief that a wrestler cannot be adequately and effectively conditioned by competition alone; hence, coaches must plan regular 'overload' circuit-training sessions using resistive principles with either commercial equipment or homemade weight and pulley systems [11, 12]. Although this method of general conditioning has been repeatedly shown in experimental conditions to be effective in increasing and maintaining whole-body strength and endurance, as well as to modestly improve aerobic capacity [11, 12], there is no assurance that these muscle characteristics can be incorporated into the specific muscular movement patterns and requirements of accomplished wrestlers. Therefore, in order to receive maximum muscular benefits from the practice sessions, the muscle contractions during the execution of select holds, throws, take downs, escapes, bridging, etc., competitors must perform at 50–100% of their maximum capacity and, if possible, for the duration of the period. These percentages must be achieved in order to acquire the necessary muscle endurance, force and power requirements to be successful.

In accordance with the concept of exercise specificity [4–6], individuals participating in Greco-Roman wrestling matches need to devote more conditioning time to the development of power in the upper-body muscles than wrestlers who are only freestyle competitors. In earlier years, the acquisition of muscular strength was considered to be the most important goal of the training program; however, we believe the attainment of muscular power and endurance are of equal or greater importance than the development of maximum force per se. To facilitate the training benefits, the duration and type of muscular contraction (concentric, eccentric or isometric) should also correspond to the duration of the period (table 1). As a general rule, the beginning wrestler should devote at least three times a week to formal muscle-resistive training procedures, although many elite wrestlers have a higher weekly frequency.

To the physiologist, the best single measurement of aerobic capacity is the measurement of maximum oxygen consumption ($\dot{V}O_2$ max). Unlike runners or longdistance skiers, there is a paucity of published data on this parameter from competitive wrestlers. One frequently cited report for an 'average' wrestler is 14.2 kcal/min [14] or approximately 3.0 liters of oxygen per minute if the respiratory exchange quotient is 1.0 [10]. Until direct, rather than indirect, methods are used to assess heat production, riding a bicycle ergometer or running on a treadmill to secure information on $\dot{V}O_2$ max will have to suffice to estimate the energy potential of wrestlers. Studies reported by Canadian researchers [2, 6–8] indicate that most international competitors average between 53 and 56 ml/kg/min, with some elite wrestlers obtaining values of 60–75 ml/kg/min. To some scientists, a wrestler with this capacity should increase the 'pace' of the match and force a wrestler with lower values into an anaerobic energy mode [13]. Interestingly, when Taylor [9] evaluated the benefits of wrestling on the aerobic capacity of adolescents, he found they had higher values than their nonwrestling counterparts, thus suggesting that the training and competitive programs had increased their ability to transport and use oxygen. Moreover, it is known that when weight loss occurs by food deprivation, a decrease in $\dot{V}O_2$ max occurs [14].

Select studies on the anaerobic capacity of elite wrestlers, measured by the concentration of lactic acid in the blood, indicated they have the capacity to produce higher levels than long-distance runners or skiers [2, 6–8]. Although the production of lactic acid does not explain the mechanism of an increased anaerobic capacity, it is a practical measure to assess this physiological characteristic [2, 10, 15]. Since the spectrum of competitive amateur wrestling matches spans a time period that ranges from 5 to 9 min (table 1), it is evident that both aerobic and anaerobic metabolic pathways are involved (table 2) and that the conditioning programs prescribed must devote a certain percentage of time to their development. This means that having wrestlers conditioned to run a marathon is unnecessary and that having them able to swim underwater for 1–3 min is worthy of consideration.

In general, athletes and coaches are obsessed with the amount of time needed for conditioning purposes, whether it is for a one-mile run (4–5 min), a marathon event (2.5–3.5 h), or a wrestling match (5–9 min). With regard to amateur wrestling in the USA and the existing practices followed at tournaments and championships, in most circumstances the wrestler has only one match per event, and it is unlikely that any wrestler will be required to compete more than three times in a given day, although the rules allow five times per day. In accordance with the concept of the specificity of training [4–6], the wrestler should train so that he can successfully complete 3–5 matches in a given day. Hence, the wrestler does not need to devote 4–6 h per day for conditioning purposes.

Nutritional Considerations

Despite the voluminous amount of material that has been written in the United States about the health practices and nutritional status of amateur wrestlers, the issues of the past are still present today, at least in this

Table 3. The estimated minimal number of calories required by adolescent scholastic wrestlers

(a)   Basal caloric requirements

Body weight kg	n	Height cm	Age years	Surface area m²	Daily minimal calories, kcal
44.5	30	159	15.0	1.42	1,544
48.6	34	164	15.4	1.54	1,674
52.3	50	167	15.5	1.60	1,728
55.9	46	169	15.9	1.66	1,781
59.1	54	171	15.8	1.70	1,824
62.3	46	173	15.8	1.78	1,910
65.9	46	173	15.9	1.82	1,952
70.4	46	178	16.1	1.88	2,017
75.0	39	178	16.3	1.94	2,081
79.5	20	180	16.5	1.98	2,100
84.1	16	179	16.5	2.08	2,206

(b)   To estimate the number of additional kcal required for wrestling practice, multiply the body weight by 11.26 kcal/h/kg

(c)   To estimate the number of additional kcal required for school activities, multiply by 1.50 kcal/h/kg

(d)   This table is a modification from a similar one published by Tipton [18] and the calculations were made using surface area, age, and metabolic data mentioned by the American College of Sports Medicine [38] and the caloric cost data listed in Brozek et al. [40].

country [16–18]. The concerns about diet, 'making weight', hazardous health practices, etc., continue to exist because wrestlers generally believe they can compete successfully in lower-weight classes with minimal impairment in performance and because they accept the premise that losing weight is an integral part of being a wrestler [16–18].

Although Buskirk [19] indicated that wrestlers need a balanced diet ranging from 3,000 to 5,000 kcal per day for normal growth and maintenance purposes, data by Short and Short [20] indicate that for collegiate wrestlers this value can range from 412 to 14,962 kcal per day, depending on the circumstances. There are many unofficial and unpublished reports that suggest the lower limit is zero. Many years ago we used age and surface area data to calculate the number of calories to maintain the basal metabolic rate of scholastic wrestlers in the State of Iowa and found a range from 1,544 kcal for a 15-year-old student weighing 44.5 kg to 2,100 kcal for a 16.5-year-old wrestler weighing 79.5 kg [14] (table 3).

Insights on the dietary and weight-loss practices of scholastic, collegiate, and international wresters have been noted by several investigators [14, 16–18, 20–23] with international-class wrestlers exhibiting fewer fluctuations [15]. The primary result was that weight loss was a common experience that occurred from 7 to more than 15 times per competitive season. Moreover, it was not uncommon for successful wrestlers to 'cut weight' 100 or more times during their careers. Unfortunately, the weight was seldom lost in a gradual or systematic manner; rather, it was lost over a period of 1–3 days especially if tournaments were involved. Contrary to expectations, neither the coach nor the parent was the source of information on how to 'make weight'; rather, it was the other wrestlers [14, 17, 18]. Regardless of the level of competition, weight is lost predominately by dehydration using thermal, exercise, and fluid deprivation methods [14, 17, 18, 21, 22]. Although the use of diuretics, laxatives and vomiting have been cited as methods to lose weight [18, 22, 23], the frequency of their use is difficult to quantify. The amount of body mass lost to 'make weight' generally ranges from 2 to 6 kg, with the loss being essentially fluids and electrolytes rather than fatty tissue [17, 18]. In fact, calculations show that a wrestler using 3 liters of oxygen per minute would need to exercise for more than 8 h to oxidize 1 kg of fatty tissue [18].

Studies conducted on interscholastic wrestlers by Oppliger and his associates indicated that approximately 2% exhibited bulimic behavior [23]. This percentage is higher than for male adolescents who are nonwrestlers and suggests that the emphasis on weight loss and on 'making weight' may cause serious medical problems for more than a 'few' wrestlers. Individuals who fluctuate in their weekly weight loss are labeled 'cyclers', and preliminary results with this population indicate that the process will decrease the resting metabolic rate [24]. Whether this is a short- or long-term effect is unknown.

The process of fluid restriction coupled with thermal and exercise dehydration increases the water and urea losses from the body and accentuates the problems of impaired thermoregulation and aerobic performance even when 2–3% of the body weight is lost [14, 25]. However, this does not appear to be the case for muscular power [26].

When attempts have been made to ascertain the dietary preferences and practices of collegiate wrestlers, it appears they consume less protein and fewer carbohydrates with more fat than recommended by authorities

[16]. In addition, these wrestlers did not consume adequate amounts of vitamins (A, B_6, thiamine and C) or sufficient amounts of iron, zinc, and magnesium [16]. Although many wrestlers believe they need to 'carbohydrate load' before competition, the calculated energy requirements for an 'average match' do not warrant the continuation of this practice. However, this conclusion could be modified for situations when two or more matches have been scheduled for one or more days. Under these conditions it would be advantageous to consume a liquid carbohydrate diet that contained between 200 and 300 kcal several hours before the second match. An alternative approach to the liquid carbohydrate drink has been published by the Canadian scientists [2] and includes combinations of tomato juice, orange juice, skim milk, honey, milk chocolate, bread, etc., depending on the amount of time between matches [2, p. 28].

In addition to information about fiber types, areas and their enzyme activities [6–8], Canadian scientists have acquired information on muscle glycogen concentrations [28]. Under normal conditions, elite wrestlers have adequate amounts of glycogen for the energy requirements of a single match. However, when they lose 8% of their body weight over a 4-day period by food deprivation and fluid restriction methods and 'work out' for 2 days, they will reduce the concentration of glycogen in the muscles by approximately 50%. Consequently, if wrestlers are not consuming normal amounts of carbohydrates in their daily diets (58%) while exercising, and if they are depriving themselves of fluids to 'make weight', they can expect a decrease in performance. This situation occurs because the glycogen levels in the liver and muscles are diminished, and the glucose concentration in the blood is inadequate to meet the needs of the central nervous system [14].

Pioneer studies by Costill and associates [29–31] have demonstrated that glycogen replenishment is not complete in the muscles or liver within the time period allowed between weigh ins. This fact also argues for the use of liquid carbohydrate diets. We favor drinks with no more than an 8% carbohydrate concentration to promote gastric emptying and the intestinal absorption of glucose, water, and electrolytes [32]. However, some individuals exhibit an insulin surge [30, 31] which is to the disadvantage of the wrestlers because they become hypoglycemic and experience a breakdown of liver glycogen. This occurs because the insulin causes the glucose, which is released from the liver into the blood, to enter the skeletal muscles. Subsequently, liver glycogen is broken down before the individual starts exercising. Since this is not a universal effect with all individuals, it is important to know whether an individual is a responder or a nonresponder to carbohydrate ingestion. Solutions containing high concentrations of fats or proteins are not recommended as precompetition or pretournament drinks because they delay gastric emptying and because fatty acid utilization as a substrate decreases when the energy requirements approach 70% $\dot{V}O_2$ max [10]. Our knowledge of the time course of electrolyte replacement in tissues is not complete. It is possible that with depletion via fluid deprivation and sweating it may require as much time as glycogen replenishment for tissue equilibriums to be restored. The end result could be an increased susceptibility of the muscular tissues to fatigue.

Minimal Weights for Wrestlers

Since 1968 an attempt has been made to have athletic and medical authorities regulate scholastic wrestling in the various states within the United States in order to implement a mandatory plan which would govern minimal wrestling weights [33, 34]. The rationale is that current procedures used by scholastic wrestlers for certification and competition for a given weight class are not (a) conducive to growth and development; (b) do not enhance learning; and (c) do not promote desirable health practices [14, 17, 18]. In fact, one County Medical Society in the State of Iowa considered the practices followed by wrestlers to 'make weight' to be 'hazardous to their health' [33].

Research published during the past decades has clearly demonstrated that it is possible to use validated anthropometrical methods to predict a body weight that has between 5 and 7% body fat [35–37]. It is important to note that the concept of a minimal body weight has been included in a position statement published by the American College of Sports Medicine [38] and that many elite scholastic wrestlers in the State of Iowa are participating in tournaments with body fat percentages that are lower than 5% [36].

For years, many communities in the United States have made attempts to initiate voluntary programs to establish minimal weights for scholastic wrestlers, but none have been successful. However, this situation may change because after careful planning and community involvement, medical, scientific and educational leaders in the State of Wisconsin have implemented a mandatory minimal weight measurement program for certification purposes for 1991–1992. To date, unpublished reports indicate the program has been accepted without major problems from wrestlers,

Table 4. Skin fold equation used to predict minimal wrestling weights in the state of Wisconsin

(a)	Selecting the skin folds used by Lohman et al. [37]
	S1–triceps
	S2–subscapular
	S7–abdominal

(b)	Determining body density
	$BD = (1.0982 - [0.000815 \times \{S1 + S2 + S7\}] + [0.00000084 \times \{(S1 + S2 + S7)^2\}])$

(c)	Determining body fat by the equation of Brozek et al. [40]
	$\text{Percent fat} = \left[\dfrac{4.570 - 4.142}{BD}\right] \times 100$

coaches, parents or administrators. We predict that after 1992 other states will use the Wisconsin program as a model for their wrestlers. Since preliminary findings suggest that the equation shown in table 4 is more accurate for Caucasian wrestlers than for those of Hispanic origin [39], more research will be necessary with students from different ethnic backgrounds. In some states this consideration will not be a major issue. In others, like Arizona, the problem will have to be resolved before mandatory implementation can occur.

Several years ago the monthly weight allowance provided for scholastic wrestlers in the United States was discontinued. However, we believe the concept should be reexamined with the following points in mind.

First, growth does occur in adolescent wrestlers, and that fact should not be ignored by the administration of wrestling programs. Second, when adolescent wrestlers 'make weight', the youngest and the lightest individuals lose the highest percentage of their body weight, which makes them more vulnerable for the physiological consequences of weight loss [17]. Third, a fixed monthly weight allowance is an advantage for the lower-weight classes and a disadvantage for the higher-weight classes. Therefore, we recommend that weight allowances be made with regard to a change in height and, as a starting point, suggest an increase of 0.45 kg weight for every 2.54 cm change in height [18].

Lastly, the principles used with scholastic wrestlers (table 4) to predict a minimal weight can be followed easily with collegiate and international competitors. It is our hope that at future conferences we will hear that this possibility has become a reality.

Conclusion

Amateur wrestling is an ancient sport with a proud and impressive tradition. Although the physiological attributes needed to be a successful wrestler are the same as during the time of Jacob, the changes in the duration of a given period indicate that the energy requirements are both aerobic and anaerobic in nature. Consequently, in accordance with the concept of exercise specificity, wrestlers must train using the movement patterns and the biochemical pathways that are unique to the sport.

The nutritional knowledge and practices of scholastic and collegiate wrestlers leave much to be desired, and it is likely that the same statement is appropriate for those on the international circuit. Contrary to facts, wrestlers at any level believe they can 'make weight' by food and fluid derivation coupled with dehydration with no consequence to their performance capacity or their ability to regulate in the heat. In many circumstances, the primary source of nutritional information comes from the other wrestlers, regardless of the intention and educational background of the coach or medical advisor. Since bulimic behavior is being reported for adolescent scholastic wrestlers, nutritional and medical authorities must work closer with coaches and competitors to prevent and to decrease this condition.

Research conducted with scholastic wrestlers indicates that it is possible to predict a minimal wrestling weight by anthropometrical methods that could minimize the problems of certification and competition at weights that have less than 5–7% fat. In addition, the research conducted by the Canadian Amateur Association during this past decade is an excellent program model for other national organizations to emulate if the fitness and nutritional aspects of the sport are to progress to the heights equal to its heritage.

References

1   King JA: Wrestling; in Larson LA (ed): Encyclopedia of Sport Sciences and Medicine. New York, McMillan, 1971, pp 422–424.
2   Sharratt CT: Wrestling profile. Clin Sports Med 1984;3:273–289.
3   Marrin D, Sharratt MT, Taylor AW: A Scientific Approach to Amateur Wrestling. Canadian Amateur Wrestling Association, 1980.
4   Henry FM: Coordination and motor learning. Proc Coll Phys Ed Assoc 1956:68–75.

5   Scheuer T, Tipton CM: Cardiovascular adaptations to physical training. Annu Rev Physiol 1977;39:221–251.

6   Taylor AW, Brassard L, Proteau L, et al: A physiological profile of Canadian Greco-Roman wrestlers, Can J Appl Sport Sci 1979;4:131–134.

7   Sharratt CT, Banister EW, Mekjanic I, et al: Physiological and Physical Status of Freestyle and Greco-Roman Junior World Wrestling Champions. University of Waterloo and Simon Fraser University, 1981, pp 1–8.

8   Sharratt CT, Taylor AW, Song TMK: A physiological profile of elite Canadian freestyle wrestlers. Can J Appl Sport Sci 1986;11:100–105.

9   Taylor AW: Physiological effects of wrestling in adolescents and teenagers. J Sport Med 1975;3:76–84.

10 Astrand P-O, Rodahl K: Textbook of Work Physiology. New York, McGraw Hill, 1986, pp 1–723.

11 Gettman LR, Ayres JJ, Pollock ML, et al: The effect of circuit weight training on strength cardiorespiratory function and body composition. Med Sci Sports 1976;10: 171–176.

12 Gettman LR, Ward P, Hagman RD: A comparison of combined running and weight training with circuit weight training. Med Sci Sports Exerc 1982;14:229–234.

13 Cunningham DA: Physiology of Wrestling. Canadian Wrestler 1980;4:5.

14 Tipton CM: Consequences of rapid weight loss; in Haskell W, Scala J, Whitman J (eds). Palo Alto, Bull, 1982, pp 176–192.

15 Sharratt M: A systematic application of science to sport. The Canadian Amateur Wrestling Association (CAWA) model. Coaching Sci Update 1984;31–34.

16 Steen SN, McKinney S: Nutritional assessment of college wrestlers. Phys Sportsmed 1986;14(11):101–116.

17 Tipton CM, Tcheng T-K: Iowa wrestling study: Weight loss in high school students. JAMA 1970;214:1269–1274.

18 Tipton CM: Making and maintaining weight for interscholastic wrestling. Gatorade Sports Sci Exchange 1990;2:1–4.

19 Buskirk ER: Some nutritional considerations in the conditioning of athletes. Ann Rev Nutr 1981;1:319–350.

20 Short SH, Short WR: Four-year study of university athletes' dietary intake. J Am Diet Assoc 1983;82:632–645.

21 Brownell KD, Steen SN, Wilmore JH: Weight regulation practices in athletes: Analysis of metabolic and health effects. Med Sci Sports Exerc 1987;19:546–556.

22 Oppliger RA, Landry GL, Foster SW, et al: Bulimic behaviors among interscholastic wrestlers: A statewise survey. In press.

23 Woods ER, Wilson CD, Masland RP: Weight control methods in high school wrestlers. J Adolesc Health Care 1988;9:394–397.

24 Steen SN, Oppliger RA, Brownell KD: Metabolic effects of repeated weight loss and regain in adolescent wrestlers. JAMA 1988;260:47–50.

25 Gisolfi CV, Tipton CM: How to 'make weight': Rubberized suit or the first law of thermodynamics. Schl Wrestl News 1974;10:6–8.

26 Horswill CA: Does rapid weight loss by dehydration adversely affect high-power performance? Gatorade Sports Sci Exchange 1991;3:1–4.

27 Sharratt MT: Physiology related to wrestling; in Marrin D, Sharratt MT, Taylor AW (eds): A Scientific Approach to Amateur Wrestling. Canadian Amateur Wrestling Association, 1980, pp 9–12.

28   Houston ME, Marrin DA, Green JH, et al: The effect of rapid weight loss on phys-
      iological functions in wrestlers. Phys Sportsmed 1981;9(11):73–78.
29   Costill DL, Saltin B: Muscle glycogen and electrolytes following exercise and thermal
      dehydration; in Howald H, Poortmans JR (eds): Metabolic Adaptation to Prolonged
      Physical Exercise. Basel, 1975, pp 352–360.
30   Costill DL, Miller JM: Nutrition for endurance sport: Carbohydrate and fluid bal-
      ance. Int J Sports Med 1980;1:2–14.
31   Costill DL, Coyle E, Dalsky G, et al: Effects of elevated plasma FFA and insulin on
      muscle glycogen usage during exercise. J Appl Physiol 1977;43:695–699.
32   Gisolfi CV: Exercise, intestinal absorption, and rehydration. Gatorade Sport Sci
      Exchange 1991;1:1–10.
33   Tipton CM, Tcheng T-K, Paul WD: Evaluation of the Hall method for determining
      minimum wrestling weights. J Iowa Med Assoc 1969;59:571–574.
34   Tcheng T-K, Tipton CM: Iowa wrestling study: Anthropometric measurements and
      the prediction of a 'minimal' body weight for high school wrestlers. Med Sci Sports
      1973;1:1–10.
35   Oppliger RA, Tipton CM: Weight prediction equation tested and available. Iowa
      Med 1985;75:449–453.
36   Tipton CM, Oppliger RA: The Iowa wrestling study: Lessons for physicians. Iowa
      Med 1984;74:381–386.
37   Thorland WG, Tipton CM, Lohman TG, et al: Midwest wrestling study: Prediction
      of minimal weight for high school wrestlers. Med Sci Sports Exerc 1991;23:1102–
      1111.
38   American College of Sports Medicine: Position stand on weight loss in wrestlers.
      Med Sci Sports 1976;8:xi-xiii.
39   Roby FB, Kempema JM, Lohman TG, et al: Can the same equation be used to
      predict minimal wrestling weight in Hispanic and non-Hispanic wrestlers? Med Sci
      Sports Exerc 1991;23:S29.
40   Brozek J, Grande F, Anderson J, et al: Densiometric analysis of body composition:
      Revision of some quantitative assumptions. Ann NY Acad Sci 1963;110:113–140.

Charles M. Tipton, PhD, Department of Exercise and Sport Sciences,
Ina E. Gittings Building, The University of Arizona, Tucson, AZ 85721 (USA)

Simopoulos AP, Pavlou KN (eds): Nutrition and Fitness for Athletes.
World Rev Nutr Diet. Basel, Karger, 1993, vol 71, pp 97–114

Jockeys and Their Practices in South Africa

D. Labadarios [a], *J. Kotze* [a], *D. Momberg* [a], *T.J.v.W. Kotze* [b]

[a] MRC Metabolic Research Group of the Department of Human Nutrition,
Tygerberg Hospital, University of Stellenbosch;
[b] MRC Institute for Biostatistics, Tygerberg, South Africa

Introduction

The weight a horse must carry in a race is determined in accordance
with the rules of the South African Jockey Club; the minimum weight is
defined as not less than 46 kg in a fixed weight race, or 48 kg in a handicap
race. It is, therefore, mandatory for the jockeys to attain or maintain as low
a body weight as possible so as to secure sufficient rides and/or wins. Arti-
cles in lay magazines indicate that, over the years, jockeys have developed
a number of techniques for reducing their body weight quickly prior to a
race and according to its specified weight requirements. These techniques
may include, singly or in combination, restriction of food and fluid intake,
fasting, exercise, the use of saunas, self-medication and hot baths. The
prevalence of these practices and their health implications, if any, are
unknown, since there is practically no available information in the scientific literature regarding any aspects of the lifestyle or general health of
jockeys. Only one study in the USA remarks on the excellent performance
of jockeys in cardiovascular endurance testing [1] and a local unpublished
study by students on a small number of Jockeys [2], highlights dehydration
as a means of body weight control.

The aims of this descriptive study have been to define the extent of the
aforementioned practices and assess the nutritional status of jockeys with
view to deciding on the practicality of existing regulations governing body
weight for horse racing.

Subjects and Methods

The study had to be so designed as not to interfere with either the programme of a race meeting or the performance of the jockeys in it. The information obtained about the activity patterns of jockeys, prior to the start of the study, indicated that their activities during the day of a race meeting differed substantially from those of other days. Jockeys had to be studied, therefore, under both conditions, so as to distinguish the short-term events transpiring the day prior to and on the day of a race meeting and the longer-term ones associated with nonracing days. Furthermore, the limited free time jockeys have during a race meeting (5 min in between races) necessitated that information regarding their established routine was best obtained at a time when they would be at relative ease and were not preparing for a race meeting. For these reasons, jockeys were studied on a Monday (baseline) since no races are held on this day; the same jockeys selected on a Monday were then studied on two separate race meeting days over the period of the ensuing week. This design ensured a high probability of collecting data representative of the jockeys' routine; any additional points of measurement would have resulted in little additional information at a logistically unacceptable expense. For practical reasons, no more than 13 and usually 10 jockeys were studied each week.

In this study, nutritional status was assessed by clinical, anthropometric, dietary and biochemical means.

Ninety-three of the maximum of 102 qualified jockeys were studied, representing 91% response. Apprentice jockeys were not studied. The 9 jockeys that did not participate in the study were either suspended, retired or overseas at the time of the investigation. Only one jockey refused to take part in the study, because he felt he had no problems with his body weight.

The study was approved by the Ethics Committee of the University of Stellenbosch. Informed consent was obtained from all jockeys and explanatory lectures were given each week regarding the week's programme and activities. All jockeys were assured of the confidentiality of the information they would provide and were asked not to deviate from the established pattern of their practices. On the whole, the cooperation received from the jockeys was excellent.

Results

Biographic-Demographic Data

All the jockeys except one (an Indian) were Caucasian. The average age of jockeys is 27.8 years and their age distribution is shown in table 1.

The majority of jockeys (63%) have an academic qualification of standard eight and 29% standard ten. Sixty-three percent of the jockeys are married, of whom 24% are married for the second or third time. Thirty-six percent are either divorced or separated. In terms of interpersonal relationships, 21% of jockeys, in their own opinion, find it difficult to make

Table 1. Age distribution of jockeys	
Age, years	Number of jockeys
19–20	5
21–25	33
26–30	34
31–35	11
36–40	4
41–45	4
46–55	2

Table 2. Percentage of jockeys reporting one or more medical problem(s)

Body system involved	Percentage of jockeys
Musculoskeletal	25.0
Gastrointestinal	13.0
Renal	5.0
Dermatological	5.0
Psychiatric	4.0
ENT	4.0
Allergy	3.0
Genital (warts)	1.0
Ophthalmological	1.0
Gout	1.0
Cardiovascular (murmur)	1.0

friends, 14% lose their friends easily and 15% mistrust their friends. Interestingly, 18% of jockeys do not see fellow jockeys as friends and 10% mistrust them. In social activities, only 62% of jockeys feel relaxed with the rest, feeling either uncomfortable or, to a lesser extent, anxious. Less than 6% of jockeys abused habit-forming drugs or alcohol, 7% had consulted a psychiatrist and 2 jockeys had attempted suicide at some stage during their childhood.

Medical

At baseline, 53% of jockeys gave a history of an active medical problem. In 2 cases the problem was of sufficient severity to advise retirement. Most of the jockeys had either had medical treatment, were under medical treatment or were on symptomatic self-treatment. The body systems involved are shown in table 2. The majority of complaints involved the musculoskeletal, gastrointestinal and, to a lesser extent, renal, dermatological, psychiatric and ENT systems. Orthopaedic complaints were related to racing accidents; renal complaints were related to renal stones or haematuria during or after a race meeting. Of the 12 jockeys with gastrointestinal complaints, 9 had medically diagnosed duodenal ulceration, 1 had recurrent vomiting during or after the races, 1 gave a symptomatology compatible with duodenal ulceration, and 1 of haemorrhoids. In addition and

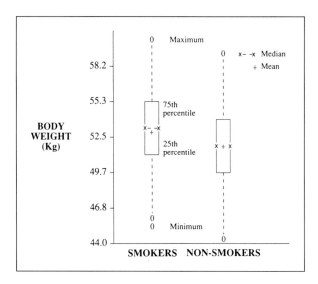

Fig. 1. The weight of jockeys in relation to smoking habits.

following the clinical examination, 4 jockeys were found to have physiologic heart murmurs and 1 hypertension.

Personal Habits

Seventy-five percent of jockeys smoke an average of 17 cigarettes daily. A fifth of them smoke more on the days of race meetings, with the rest smoking either the same or slightly less. On average, they start smoking at the age of 16 and some as early as 10–12 years of age. Although no significant difference in the weights of smokers and non-smokers was found in this study (fig. 1), a significant proportion (58%) of jockeys use smoking as a means of controlling their body weight. On average, smokers tended to be heavier than non-smokers, but it could not be ascertained whether the weight of the smokers would increase if they were to stop smoking, a claim made by many of the jockeys. Tradition also plays an important role in their decision to start smoking (38%) as does their environment. Forty-four percent of the jockeys, for instance, smoke because 'the other jockeys smoke' and 34% smoke in order 'to combat stress'. A lesser percentage smoke for other reasons, including family and social influences and 'in order to improve their concentration'.

Exercise is a prominent feature in the jockeys' lifestyle and on average they exercise 2–3 times per week for 1.5 h. The most popular forms of exercise are running, walking, swimming and squash. Eighty-two percent of the jockeys are of the opinion that exercise is very important in controlling their weight, especially when exercise is combined with the wearing of layers of heavy clothing and/or plastic suits ('sweat suits'). In addition, apart from riding on the day of the race meeting, jockeys also ride (riding work) 5–6 mornings a week for an average of 4 h, as part of the horse's training programme.

Apart from a small minority (< 5%) who had used or are using cannabis occasionally, none of the jockeys use any other habit-forming drugs. However, there exists widespread use of drugs relating to racing performance and preparation for racing; for instance, approximately 1 of 5 jockeys use non-steroidal anti-inflammatory drugs (NSAIDs) including aspirin as well as central nervous system stimulants such as caffeine (see also section on body weight).

Anthropometric

Body Weight. In general, jockeys are not 'light babies' at birth. Their average birth weight is 3.0 kg and only 20% of them have a birth weight of less than 2.5 kg. Furthermore, no significant correlation was found between their weight at birth and their present weight, although there was a tendency towards lighter jockeys being lighter babies.

Jockeys enter the Academy with an average body weight of 33 kg and during their 5-year apprenticeship they gain an average of 14.5 kg. The average body weight of jockeys in South Africa at present is 52.9 kg; the lightest jockey is 44 and the heaviest 60 kg. The distribution of jockeys according to body weight at the time of the first interview is shown in table 3. Only 8 jockeys (fig. 2) have a weight equal to or less than the minimum weight of 48 kg, the currently used regulatory weight for a handicap race. Furthermore, only 5 of the jockeys studied are above their ideal weight (fig. 3) for age and height.

Sixty-six percent of the jockeys are unable, at present, to maintain a constant body weight and 64% are experiencing increasing difficulty in maintaining an appropriate body weight for a given race. This difficulty first manifested itself at between 19 and 20 years of age and in 73% of cases gradually increased in magnitude with time.

Over the year preceding the study, 75% of jockeys had lost a maximum of 2–6 kg for a race, 5% had lost more than 6 kg and some had lost as

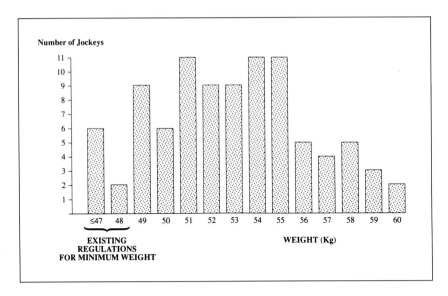

Fig. 2. Body weight distribution of jockeys in relation to existing regulations for minimum weight for horse racing.

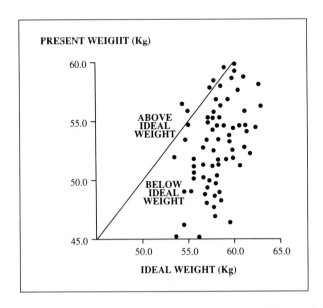

Fig. 3. The weight of jockeys in relation to their ideal weight for age and height.

Table 3. Distribution of jockeys according to weight at baseline

Weight, kg	Number of jockeys
≤47.0	6
48.0	2
49.0	9
50.0	6
51.0	11
52.0	9
53.0	9
54.0	11
55.0	11
56.0	5
57.0	4
58.0	5
59.0	3
60.0	2

Table 4. Maximum weight lost for a race over the year preceding the study

Weight lost, kg	Total	
	number of jockeys	%
0	2	2.1
0.1–2.0	16	17.2
2.1–4.0	49	52.7
4.1–6.0	21	22.6
6.1–8.5	5	5.4
Total	93	100.0

much as 8.5 kg (table 4). On a weekly basis and over the same period, 58% of jockeys lost up to 2 kg for a race meeting and 33% of them lost between 2 and 4 kg. A lesser percentage (3%) lost 4–6 kg per week (table 5). In table 5, a comparison is made between the claimed weight loss by jockeys and the loss actually measured during the study. The discrepancy in the percentages reporting no loss of weight is probably due to the jockeys not reporting, or not being aware of, small losses in body weight. It appears, therefore, that jockeys tended to underestimate their loss of body weight, a tendency that appears in the rest of the table for the two race meetings. Less than 4% of the jockeys thought that their body weight loss was in any way related to any previous accidents they may have had. The average time over which the reported body weight loss occurred was 17 and 12 h, respectively, for the two race meetings. This time, although usually short in the majority of cases, could be shorter or longer, depending on circumstances and on how much weight had to be lost.

In order to combat their 'weight problem', 77% of the jockeys use 'dieting' (short-term reduction or cessation of food and fluid intake), 70% use saunas, 80% exercise (with 48% wearing 'sweat clothes' while they

Table 5. Percentage of jockeys claiming a weekly weight loss in relation to the actual weight loss measured

Weight loss kg	Weekly weight loss claimed at baseline (non-race day)	First race meeting weight loss		Second race meeting weight loss	
		claimed	measured[1]	claimed	measured[2]
0	5.4	34.9	6.2	45.5	9.0
0.1–2.0	58.2	45.3	50.0	40.0	59.0
2.1–4.0	33.4	19.8	26.3	14.5	21.5
4.1–≥6	3.0	–	5.0	–	3.5

[1] 12.5% of jockeys gained 0–1.8 kg.
[2] 7.0% of jockeys gained 0–1.8 kg.

Table 6. Percentage of jockeys using drugs in their attempt to control their body weight

Drug used	Claimed use at baseline	Claimed use for first race meeting	Confirmatory/ suggestive urine test[1]	Claimed use for second race meeting	Confirmatory/ suggestive urine test[1]
Diuretics	70.0	30.0	53.0	13.0	55.0
Laxatives	27.0	6.0		2.0	
Appetite suppressants	48.0	10.0	12.0	11.0	16.0

[1] The higher percentage of confirmatory/suggestive urine test for diuretics and laxatives combined may be due to the fact that the screening test used was also positive for the presence of drugs other than appetite suppressants.

exercise) and 27% use hot baths. In addition, 70% use diuretics, 27% laxatives and 48% appetite suppressants. Table 6 summarises the percentage of jockeys who claimed to have used these drugs at baseline together with the percentage of jockeys who claimed to have used these drugs in preparation for the two race meetings. The percentage of confirmatory/suggestive urine tests is also given for comparison. The overestimation of drug usage at baseline is due to the fact that the main determinant of drug use is how 'light a ride' a jockey has for a particular race meeting. The similarity in the percentages between reported use and use as detected in the urine suggests that drug abuse for body weight control is widespread among jockeys. Jockeys abuse these drugs, particularly diuretics, despite their belief that they can be 'harmful' to the kidneys and muscles. Although 92% also know that diuretics cause loss of body water, electrolytes and minerals, 26% erroneously believe that they also cause loss of body fat.

The overwhelming majority (91%) of jockeys use these methods because they consider them to be effective; they have arrived at the methods of their choice by self-experimentation. They obtained their information mainly from other jockeys and adapted it in the development of a method that 'suits them'. Only 3% of the jockeys had sought medical advice in this regard. The use of these methods, however, is not without undesirable side effects of which irritability (23%), tiredness (20%) and dizziness (7%) were the most commonly reported on both race meetings.

Height. The length of jockeys at birth, although asked for, could not be obtained in the majority of cases. Data obtained from jockeys who could remember their Academy entry and graduation heights indicate that during their 5-year stay in the Academy their height increased by 15.20 cm. At present, the average height of jockeys is 160.9 cm, the shortest jockey having a height of 144.1 and the tallest 174.0 cm.

Body Fat. The average percentage body fat determined by measuring skinfold thickness at four body sites was 11%. This is well below the average for a normal population and compares favourably with that of some other athletes. In fact, 28% of jockeys have a percentage body fat of less than 10% and only 5% of them have more than 15% body fat.

Dietary Intake

The dietary intake was assessed by a dietary history, a 7-day food record and a 24-hour recall on the morning of the two separate race meetings. The mean intake of energy together with the mean intake of the nutrients that fall below the recommended daily intake is presented in table 7. The nutrient intake from vitamin and other supplements is not included. When considered as a group, jockeys diminish their dietary intake in the short term, namely the day preceding a race meeting. However, in the longer-term, they tend to make up on non-race days, as shown by the increased intake from the 7-day record and dietary history data. The close agreement of the data between the two separate 24-hour recall points more than likely depicts the established pattern and the sometimes severe discipline jockeys exercise in the short-term reduction of their body

Table 7. Mean dietary intake[1] of energy and the intake of nutrients that fall below recommended daily intakes[2]

Energy/ nutritent	RDA[3]	1st race meeting 24-hour recall	2nd race meeting 24-hour recall	Seven-day record	Dietary history
Energy, kJ	–	7,400	7,400	8,100	10,400
Fiber, g	20	11	13	11	17
Calcium, mg	800	454	470	460	760
Magnesium, mg	350	220	216	235	315
Zinc, mg	15	9.3	7.8	10	12.4
Thiamin, mg	1.4	0.96	0.97	0.98	1.30
Riboflavin, mg	1.6	1.17	1.08	1.23	1.78
Vitamin B_6, mg	2.2	1.09	0.97	1.23	1.67
Folic acid, µg	400	169	167	192	273
Pantothenic acid, mg	4.7	3.6	3.2	3.8	5.3

[1] Rounded to nearest decimal or 100 kJ as applicable.
[2] The nutrient intake from vitamin and other supplements is not included.
[3] Recommended dietary allowance.

Table 8. Mean percentage contribution of macronutrients, sugar and alcohol to the total daily energy intake, of jockeys in relation to that recommended

Nutrient	Mean contribution to total daily energy intake, %				
	1st race day 24-hour recall	2nd race day 24-hour recall	seven-day record	dietary history	currently recommended
Protein	15.8	12.8	15.2	15.2	15
Total fat	35.8	31.3	33.9	34.7	< 30
Saturated fat	12.5	10.5	11.5	12.1	< 10
Polyunsaturated fat	8.3	7.9	7.8	7.7	not greater than 10
Total CHO	43.3	50.1	43.4	43.2	> 55[1]
Sugar[2]	16.6	20.7	16.6	16.9	–
Alcohol	0.7	0.83	7.9	0.94	–

[1] With emphasis on greater use of whole grains, fruits and vegetables.
[2] In very active people ten percent of the energy intake can be in the form of refined carbohydrates.

Table 9. Percentage of jockeys with low nutrient blood value at baseline

Nutrient	% with low values
Folic acid	44.0
Magnesium	34.0
Iron	26.0
Riboflavin	25.0
Thiamin	12.0
Zinc	12.0

weight. Approximately one-third of the jockeys have to reduce their dietary intake in the short-term to less than 1,200 kcal (5,000 kJ) (the proposed minimum at which good nutrition can still be achieved [3]) and 17% to less than 1,000 kcal (4,200 kJ). In the longer term, however, these low-energy diets are only followed by 11% of the jockeys; 9% consume less than 1,200 kcal (5,000 kJ) and 2 jockeys consume less than 1,000 kcal (4,200 kJ) according to the information obtained from the 7-day record and dietary history data.

The percentage contribution of macronutrients, sugar and alcohol to the daily energy intake is presented in table 8. In view of the accepted important role of carbohydrates in physical performance and endurance, it is of interest to note that on average the carbohydrate contribution to the energy intake is considerably below the recommended level. In fact, in only 2 jockeys did carbohydrate intake contribute more than 55% of the daily energy intake. By contrast, more than 40% of jockeys had a protein contribution to energy intake greater than 15%; similarly, percentages for fat intake exceeded that recommended, and more than 70% of jockeys had a saturated fat contribution to energy intake greater than 10%.

Biochemical

The percentage of jockeys with low nutrient blood values at baseline is shown in table 9. Although the average dietary intake of some of these nutrients was also low (table 7), caution should be exercised in associating the two; however, inadequate dietary intake can be considered as a contributory factor to the low blood nutrient levels found.

The baseline, pre- and intra-race values for urinary specific gravity and blood parameters related to haemoconcentration, namely total pro-

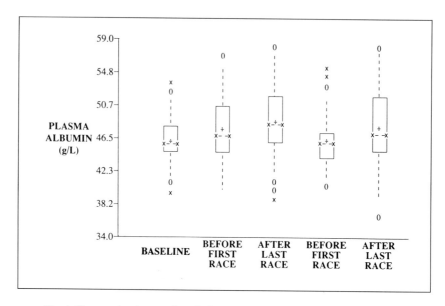

Fig. 4. Changes in plasma albumin in relation to body weight loss in jockeys.

tein, albumin, urea, creatinine and haematocrit, together with a clinical evaluation of hydration status, show that, in general, these parameters rose significantly from baseline to the time of the last race of a race meeting. Further, the rise in these values was positively and significantly related to the amount of weight lost over this period. The representative pattern of change for one of the aforementioned parameters (plasma albumin) is shown in figure 4. These data confirm the clinical findings of a state of dehydration.

Performance of Set Tasks
    The ability of jockeys, as a group, to perform certain set tasks correctly, such as memorise, recall, draw a design, identify a randomly scattered number or combined number and letter sequence as well as the accuracy of task performance, reaction and response time, show that jockeys were less able to perform these tasks correctly before the first race of a race meeting than they were at baseline. This decline in performance, for certain tasks, was significantly correlated to the amount of weight lost in preparation for the race meeting.

Discussion

The findings of this investigation show that a large percentage of jockeys practice a short-term body weight reduction programme twice weekly. Although the extent of the weight loss may vary from week to week, it nevertheless occurs throughout the year. The repetitiveness and chronicity of this short-term body weight reduction makes the jockeys unique as a group; other athletes, e.g. wrestlers in the USA, follow similar practices but on a less frequent basis and over a relatively shorter period [4].

The necessity for jockeys to lose weight for a race arises mainly from the requirements of the existing regulations governing the scale of minimum/maximum weights for a race. As far as can be ascertained from the scanty literature on the subject, this scale of weights was first developed in England in the 19th century and since then it has been occasionally adjusted on empirical rather than scientific grounds [1]. The applicability of the existing scale of weights to the conditions of today must be questioned, particularly in view of the fact that the man of today is heavier and taller than his counterpart of years ago. Evidence of this trend has been acknowledged and changes have been incorporated into the more recent tables defining ideal weight for the general population [5]. The same pattern holds true for athletes – in 1942 the percentage body fat and lean body mass of offensive/defensive linesmen in the USA was 14% and 83.1 kg, respectively [6]; in 1976 [7], the comparable values were 16.8% and 95.2 kg, an increase of 20% in body fat and 14.6% in lean body mass, respectively.

Irrespective of the limitations of the existing scale of weights, this study has shown that the overwhelming majority of jockeys have no strategy for medium- to long-term body weight control and rely almost exclusively on traditional methods of body weight control which include, singly or in combination, transient food and fluid restriction, fasting and exertional, thermal or medicinal dehydration. Water is accepted as one of the critical nutrients for any competing athlete and its absence or induced loss is not only counterproductive in terms of athletic performance, but is also potentially life threatening [8]. Although the effects of rapid weight loss by dehydration on athletic performance has not been extensively documented, muscle strength and endurance, both of great importance to jockeys, are reported to diminish when dehydration exceeds 3% loss of body weight [9]. According to these criteria, a significant proportion of jockeys are riding in a race in a suboptimal physical condition and up to 12% may be in danger of heat exhaustion. Significantly, available evidence indicates that muscle strength and endurance, both of importance to the jockey's performance, do not return to normal, 4 h after the fluid loss is restored; the jockeys' practice of dehydrating themselves for a particular race and then attempting to rehydrate themselves during the course of a race meeting, hoping that their performance will improve, is not to be recommended. Furthermore, water deprivation does not only cause heat exhaustion, but may also affect kidney function. Renal blood flow is known to be reduced during dehydration and evidence of acute episodes of renal ischaemia have been documented in wrestlers in the USA, who employ practices similar to those employed by jockeys for controlling their body weight [10]. Jockeys who concomitantly ingest NSAIDs may be at greater risk. NSAIDs have a common toxicity profile. The main mode of action of NSAIDs is reduced synthesis of prostaglandins by inhibition of prostaglandin synthetase. Renal insufficiency associated with reduced prostaglandin levels is reported to be more predictable and is observed more frequently in salt-depleted and/or dehydrated individuals [11]. The long-term significance of these findings, however, is still unknown. Furthermore, the use of NSAIDs is known to be associated with gastrointestinal toxicity and can occur in up to 20% of patients on full-dose schedules, while peptic ulceration can occur in 1–3% of the cases. The prevalence of peptic ulceration detected in this study identifies another potential health risk which warrants further investigation.

The use of drugs in sports probably dates back to 1890 when the first death of a British cyclist using ephedrine to improve his performance was recorded [12]. The extent of drug (ab)use among jockeys is not well documented. In North America in 1984, thirty-two of nearly 3,000 licensed jockeys were charged with drug and/or alcohol-related offences. In the same year in the United Kingdom, none of the 23 cited offences were drug related [13]. In South Africa, with the exception of a small percentage of jockeys who have experimented with cannabis and those who use central nervous system stimulants such as caffeine, the main classes of drugs abused by jockeys are diuretics, laxatives and appetite suppressants. The abuse of these drugs is not without danger, especially when they are abused frequently, as they appear to be, among the heavier jockeys. The potential health hazards associated with the abuse of diuretics and appetite suppressants dictate that this practice should be discontinued.

The combination of the use of diuretics with saunas may expose jockeys to additional health risks. Although saunas are generally considered to be a safe means of relaxation, their abuse, as is often found in jockeys, may be dangerous. In an already dehydrated person with reduced blood volume, as was found in this study, or in a person who becomes dehydrated because of prolonged exposure to the hot environment of the sauna, body core temperature may rise and heat stroke, heat exhaustion, muscular cramps and unconsciousness may ensue. Some cases of sudden death following the abuse of saunas have been reported, among others, in a 30-year-old jockey who died of a fatal arrhythmia after taking diuretics and using the sauna for 2 h [14]. Although no arrhythmias were diagnosed in this study, nevertheless the literature highlights the potential dangers inherent to the practices jockeys adopt to lose weight in the short-term.

The erratic pattern of eating as shown by the various percentages of them missing meals repeatedly, restricting food and fluid intake and following the binge/fast cycle may have important implications in terms of body weight management. In this study, jockeys consuming a smaller number of meals were heavier; although the reason for this observation is not clear, emerging evidence in the literature indicates that food efficiency may be increased in athletes with low body weights [15]. Certainly, animal studies indicate that repeated bouts of weight loss and weight gain increase food efficiency [16]. In these animals, weight loss occurs more slowly and weight gain more rapidly with repeated bouts of weight loss and regain. The possibility must, therefore, be considered that jockeys may find it difficult to lose and easy to gain weight because of the pattern of their 'dieting' and eating.

Pathogenic weight control, defined as weight control by restriction or cessation of food intake, binging more than twice weekly, dehydration, self-induced vomiting, and/or use of laxatives, diuretics and appetite suppressants, is not restricted to jockeys. In the USA it is also practiced, for various reasons, by a significant percentage of other athletes including wrestlers [4], ballet dancers [17], university majorettes [18], swimmers [19] and gymnasts [20]. The American Medical Association in a position statement has condemned these practices because of the inherent potential dangers to the athletes practising them [21].

In conclusion, this study has identified four main contributory factors to the practices jockeys employ for weight control, namely tradition, the existing unrealistic regulations governing the present scale of weights, lack of long-term strategy for body weight control and lack of nutrition knowledge among jockeys. It is, therefore, recommended that in the overall management of the weight control difficulties jockeys experience, these factors are taken into consideration in formulating a sound strategy for improving their overall health and lifestyle.

Conclusion

The practice of 'wasting' or 'making weight', the popularized term for losing weight quickly, is reputed to be widespread among jockeys. This descriptive study was undertaken to assess the prevalence and extent of body weight loss, the methods used to meet weight requirements for a race, the dangers, if any, inherent in these methods as well as the nutritional status of jockeys.

Ninety-three of a maximum of 102 jockeys have been studied representing a 91% response. The average age, weight and height of Jockeys in South Africa is 27.8 years (range 19–55), 52.9 kg (range 44–60) and 160.9 cm (range 144.1–174.0), respectively.

Over the year preceding the study, 75% of jockeys had lost 2-6 kg in weight in preparing for a race, while 5% of them had lost more than 6 kg and some as much as 8.5 kg. This weight loss was achieved by short-term 'dieting' (reduction or cessation in food and fluid intake) in 77% of jockeys, the use of saunas in 70%, exercise in 80% (with 18% wearing 'sweat clothes' while they exercise) and hot baths in 27%. In addition, 70% of jockeys abused diuretics, 27% laxatives and 48% appetite suppressants for their short-term body weight management.

As a group, jockeys have no strategy for long-term body weight management and rely almost exclusively on traditional methods of body weight control, which include, singly or in combination, exertional, thermal or medicinal dehydration. Their pathogenic weight control is not only associated with real and potential health hazards such as loss of strength, compromised thermoregulation, renal ischaemia and heat exhaustion but it may also, and on occasions does, adversely affect their ability to perform certain tasks adequately.

Fifty-three percent of jockeys gave a history of an active medical problem, of which the majority involved the musculoskeletal system, resulting from racing accidents, and the gastrointestinal system, mostly due to peptic ulceration. Two possible risk factors have been identified in the case of the latter, namely vulnerability to stress and ingestion of NSAIDs.

Jockeys have an erratic pattern of eating and consume a diet that is inadequate in certain nutrients, higher in fat, particularly saturated fat, and lower in carbohydrates than is recommended. Overall, their diet predisposes them to long-term health risks and is inadequate for optimal performance.

The applicability of the existing scale of weights governing minimum and maximum weight for a race is questioned and, together with tradition and lack of sound nutrition knowledge, appears instrumental to the methods jockeys employ to control their body weight.

Acknowledgements

The help received from the General Manager, Stipendiary Stewards and other Officials of The Jockey Club of South Africa and Turf Clubs in the execution of the study is acknowledged.

Thanks are due to the Departments of Dietetics, Tygerberg Hospital and Psychology, University of Stellenbosch for contributing to the composition of the research team. The MRC's Institute of Biostatistics was responsible for the statistical aspects of data analysis.

Appreciation is also expressed for the excellent co-operation received from the Management and Departments of Chemical Pathology, Haematology, Radiology, Orthopaedics and Urology of Tygerberg Hospital as well as the Department of Pharmacology, University of Stellenbosch and the Department of Radiology, University of Natal.

The excellent technical and secretarial assistance of the Personnel of the Metabolic Research Group is acknowledged.

The study would not have been made possible without the co-operation and help received from all the jockeys.

Dedicated to Mr. Demetre Bougas.

References

1   Maddox D: For jockeys, small is beautiful – but difficult. Physician Sportsmed 1976; 4:88–89.
2   Geldenhuys L, Ruler L, Fell L: Community Medicine Project: Health Hazards of Jockeys. University of Witwatersrand, Department of Community Health, MB ChB IV Student Internal Report, 1985.
3   Robinson CH, Lawler MR: Normal and Therapeutic Nutrition. New York, MacMillan, 1977, pp 205–207.
4   Tipton CM, Tcheng Tse-Kia: Iowa wrestling study: Weight loss in high school students. JAMA 1970;214:1269–1274.
5   Metropolitan Height and Weight Tables: Stat Bull 1983;64:1–9.

6 Welham WC, Behnke AR: The specific gravity of healthy men. JAMA 1942;118: 490–501.

7 Wilmore JH, Par RB, et al: Football pro's strength and CV weaknesses – charted. Physician Sportsmed 1976;4:44–45.

8 Hecker AL: Nutritional conditioning for athletic competition. Clin Sports Med 1984;3:567–582.

9 Torranin C, Smit DP, et al: The effect of acute thermal dehydration and rapid rehydration on isometric and isotonic endurance. J Sports Med Phys Fitness 1979; 19:1–9.

10 Zambraski EJ, Foster DT, et al: Iowa wrestling study: Weight loss and urinary pro-files of collegiate wrestlers. Med Sci Sports 1976;8:105–108.

11 Calabrese LH, Rooney TW: The use of nonsteroidal anti-inflammatory drugs in sports. Physician Sportsmed 1986;14:89–97.

12 Cooper DL: Drugs and the athlete. JAMA 1972;221:1007–1011.

13 Thomas S: Sport and drugs. Blood-Horse 1984;60:9481–9483.

14 Luurila OJ: Cardiac arrhythmias, sudden death and the Finnish sauna bath. Adv Cardiol, vol 25. Basel, Karger, 1978, pp 73–81.

15 Brownell KD, Steen SN, et al: Weight regulation practices in athletes: Analysis of metabolic and health effects. Med Sci Sports Exerc 1986;19:546–556.

16 Brownell KD, Greenwood RC, et al: The effects of repeated cycles of weight-loss and regain in rats. Physiol Behav 1987;38:45–64.

17 Peterson MS: Nutritional concerns for the dancer. Physician Sportsmed 1982;10: 137–143.

18 Humphries LL, Gruber JJ: Nutrition behaviour of university majorettes. Physician Sportsmed 1986;14:91–98.

19 Dummer GM, Rosen LW, et al: Pathogenic weight-control behavior of young com-petitive swimmers. Physician Sportsmed 1987;15:75–84.

20 Rosen LL, McKeag DB, et al: Pathogenic weight control behavior in female athletes. Physician Sportsmed 1986;14:79–86.

21 American College of Sports Medicine: Position stand on weight loss in wrestlers. Med Sci Sports Exerc 1976;8:xi–xiii.

Dr. Demetre Labadarios, MRC Metabolic Research Group,
Department of Human Nutrition, University of Stellenbosch, PO Box 19063,
Tygerberg 7505 (South Africa)

Special Presentations

Simopoulos AP, Pavlou KN (eds): Nutrition and Fitness for Athletes.
World Rev Nutr Diet. Basel, Karger, 1993, vol 71, pp 115–143

Exercise and Mood

Regina C. Casper

Department of Psychiatry and the Committee on Human Nutrition,
The University of Chicago Hospital, Chicago, Ill., USA

Introduction

Post-archaic hellenic (Greek) sculpture stands unrivalled in its expression of harmony between body and mind. Inspired by the agonistic ideal practiced during the Panhellenic Games, the notion of soundness of mind and body has ever since been associated with Greece's cultural and military accomplishments.

Exercise as a purely recreational activity is a fairly novel phenomenon, not surprisingly, since, until the beginning of this century, human labor and manual skills were a requirement for the production of most goods. The majority of people worked physically, while a privileged few could afford a sedentary life of leisure. With automation, the level of physical activity has been dramatically reduced for all and, hence, physical demands in the workplace and at home have also decreased.

The religious significance of the Panhellenic Games has vanished. The new Olympic games still inspire mutual tolerance, and since their revival, sports have been a privilege of athletes. Not until a growing number of health-consciousness individuals in the 1970s began to engage in regular strenuous exercise beyond their normal activities and turned into enthusiastic acolytes of exercising was the public informed about potential benefits of extra physical activity. In the last decade, a good number of studies have been conducted to explore the putative benefits of exercise and to define more precisely what is meant by the assertion of 'feeling better'.

The purpose of this chapter is to review the psychological effects, positive or negative, of vigorous physical activity or exercise: (1) on the well-being of normal populations, especially on feelings of depression or anxiety; (2) on the personality and self-concept of healthy individuals and of athletes, and (3) on clinically symptomatic populations.

These sections will be followed by a review of studies on the adverse effects of exercise. Finally, the explanatory theories examining various mechanisms possibly underlying the mood changes will be explored. The psychological effects of exercise coexist with the physical effects and are probably related to each other. Since not enough is known about this association, information on physical performance measures will be listed, when available, however space limitations do not permit a complete listing of physical measures for each case.

The Most Commonly Used Instruments for Psychological Assessment

Anxiety

The prevalence of anxiety or anxiety states associated with stress or psychological problems in the general population has been estimated to be from 2 to 4% [1]. The most commonly used instrument in exercise studies for measuring anxiety levels is the Spielberger State-Trait Anxiety Inventory (STAI) [2], a 40-item self-evaluation questionnaire which inquires into positive and negative feelings and yields separate measures of situational (state) and enduring (trait) anxiety. Items on the scale refer to the symptoms or behavior that is being rated (e.g. calm, nervous, sad).

Completion of the STAI takes about 5 min. The STAI does not inquire into physiological symptoms, such as palpitations, sweating, or agitation; these symptoms are part of the anxiety scale of the MMPI [3] and the MMPI-derived Taylor Manifest Anxiety Scale [4]. It is important to realize that affective states under normal conditions tend to be transitory and can be modified by various types of interventions.

Depression

Depressive symptoms as well as feelings of well-being have generally been measured by the Profile of Mood States (POMS) [5], a 60 (or recently, 72)-item, 10-min self-report questionnaire which provides measures of tension (anxiety), depression, anger, fatigue, confusion and vigor, additionally, friendliness and elation on the amended 72-item form. The MMPI depression subscale [3] and the Beck Depression Inventory [6] – both also self-report scales – provide more comprehensive measures of depressive symptoms.

Personality

Personality tendencies are most often assessed by the Catell 16-Factor
Personality Inventory [7], which yields 16 not entirely independent bipolar
dimensions and a second-order factor of emotionality or anxiety or the
Eysenck Personality Inventory (EPI) [8], which falls into three principal
scales, introversion, extroversion, and neuroticism.

Psychological Changes Induced by Exercise in Normal Populations

Noncompetitive exercise has become a leisure activity in Western
countries with the vast majority of people who follow a structured exercise
program reporting beneficial emotional, cognitive, or physical changes.
Doubtlessly, the enthusiasm of devoted exercisers has done the most to
create the wide-spread movement towards encouraging regular vigorous
physical exercise in the normal population. The experiences described,
which range from feeling good, relaxed, euphoric, imaginative, a sense of
accomplishment, improved self-worth, to better concentration and vivid
sensations, raise the question whether the beneficial effects might not be
merely a consequence of an intense, persistent involvement in a meaning-
ful activity which might improve emotional stability, self-sufficiency, and
conscientiousness? Alternatively, exercise could act as a stimulant and provide a heightened sense of well-being, a euphoric high, an enhanced sense
of self and greater appreciation of the surrounding world? Or exercise
might eliminate simply awareness of fear, tension, and irritability or dis-
tract from worry and dejection and in this way counteract inertia, fatigue,
depression, or confusion?

The Psychological Effects of Exercise on Anxiety and Depressive
Feelings in Normal Populations

Anxiety is an uneasy apprehension over an impending or unknown ill
or danger and far more common in women than in men. Two forms of
anxiety are distinguished. The transitory condition consists of a subjective
sense of apprehension associated with autonomic nervous system activation called state anxiety. It is this form of anxiety which has been consistently shown to decrease in response to a single bout of exercise and in
response to habitual exercise (table 1). Trait anxiety, a relatively stable and
permanent disposition to experience anxiety, is much less affected by exercise. Schwartz et al. [9] have reported cognitive elements of anxiety, lack of

Table 1. Effect of exercise on mood in healthy populations

Subjects	Number	Age	Fitness	Experimental measures	Observed effect	Ref. No.
Sedentary women	22	n.s.	physical fitness program, 4 ×/week 40 min for 20 weeks	anxiety neuroticism measures, palmar sweat test, stroke volume, heart rate	reduced anxiety, neuroticism, improvement palmar sweat test and cardiovascular function	98
Normal males	67	26–55	grip strength; physical working capacity: jogging, swimming, etc., 3 ×/week for 6 weeks	Zung Self-Rating Depression Scale	no overall effect; 'felt better', 11 depressed subjects improved	99[a, b]
High risk for MI volunteers from police	36		exercise 3 ×/week for 12 weeks vs. no exercise group	Multiple Affect Adjective Check-list, cardiovascular fitness	reduced anxiety and depression, improved physical fitness	100
Male volunteers	75	22–71	20 min treadmill, meditation, resting	STAI, heart rate	equal reduction in state anxiety in all groups; increased HR in exercisers	100
Female and male volunteers	16	25–61	exercise 3 ×/week for 10 weeks vs. sedentary controls	POMS; STAI; Retrospective Change Questionnaire	exercise reduction in tension, fatigue, confusion, state trait anxiety, greater vigor perceived, improved health and personal achievement	101[b]
Beginning and trained male and female swimmers	100	mean 22.3	swimmers vs. students attending lectures	POMS, lie scale	reduced tension, depression, anger, confusion, increased vigor, no gender effect	14
Male and female students and employees	64	mean 25	exercise 3 × 45 min. vs. hobby class	Nowlis Mood Scale; POMS Stroop Color and Word Test	exercise effect on elation, less sadness and seriousness, fatigue, better cognitive processing, no difference on POMS	102[e]
Male volunteers	30	20–30	highly trained and untrained subjects	Arithmetic Task: Stroop Color Word Task; plasma catecholamines, cortisol, acid, STAI, subjective arousal	trained subjects less arousal and earlier epinephrine peaks and higher baseline prolactin levels	95

Subjects	N	Age	Exercise program	Measures	Results	Ref.
Student volunteers	51	19–30	exercise: 90 min 5 ×/week; control: 60 min 2 ×/week for 6 weeks	Taylor Manifest Anxiety Scale, MMPI items; introversion-extroversion; subjective well-being	high exercise greater improvement than low exercise group in cardiovascular fitness; high exercise reduction in anxiety	103[d,e]
Female students	20	n.s.	high fitness vs. low fitness	memory test, subjective arousal, heart rate	high fit subjects had smaller increases in pulse rate during stress	104
Male and female	112	17–31	low and high levels of fitness, groups recording health problems and stress for 9 weeks prospectively	Life Experiences Survey Health Problems Questionnaire; Beck Depression Inventory STAI; Discomfort-Alienation Scale	low fitness subjects more health problems and more depression when life stress was high than physically fit subjects	105
Male volunteers	14	–	crossover design; 12 weeks exercise 5 ×/week; 12 normal physical activity	POMS	no mood changes	106[e]
Experienced male distance runners	7	–	80 min at 40 or 60% VO_2 max 40 min at 80% VO_2 max	POMS; plasma epinephrine norepinephrine, dopamine and lactate, enkephalins	high exercise greatest increase in catecholamines, decrease in tension at 60 and 80%	89
Male and female college students	55	mean 19	aerobic exercise 3 ×/week for 11 weeks progressive relaxation no treatment	Health and Life stress forms; Beck Depression Inventory; STAI; Hopkins Symptom Checklist	no differences in health, exercise resulted in greater reduction in depression than relaxation in high life stress groups	107
Normal adults	128	17–71	exercise 4 groups, from jogging to mild walking vs. controls 90 min 2 ×/week for 8 weeks	POMS; Internal-External Control, Marlowe-Crowne Social Desirability Scale	improvement in overall mood and confusion scale in all fitness groups greatest in mood disturbed subjects	68[c,d]
Female volunteers	40	18–22	high intensity vs. low intensity cycling for 15 min	POMS; invigorated, refreshed; uplifted; anxiety; heart rate	tension-anxiety increased in high exercise group; no effect on depression, higher exhilaration and lower heart rate in highly fit individuals	38
Male and female volunteers	101	60–83	aerobic exercise, yoga and waiting list 3 ×/week for 16 weeks	STAI; CES Depression Scale, Affect Balance Scale, Hopkins Symptom Checklist, neuropsychological functioning, finger tapping test; cardiorespiratory fitness	male exercisers reduction in depression and trait anxiety, female exercisers reduction in state anxiety, exercise and yoga groups perceived improved quality of life	13

(Table 1 continued next page.)

Table 1 (continued)

Subjects	Number	Age	Fitness	Experimental measures	Observed effect	Ref. No.
Male and female healthy volunteers	66	16–75	Tai Chi exercises before-during-after	POMS; STAI; salivary cortisol, urinary amines during Tai Chi	changes in all POMS scores during and after; increased pulse and NE; lower cortisol	108
Male and female adult volunteers; not depressed	109	18–60	high and moderate exercise 3 ×/week attention group mild exercise and waiting list; post-training and 3-month follow-up	POMS; perceived coping ability; physical well-being	high exercise greatest improvement in aerobic fitness; no change in mood except reduced anxiety/tension and confusion in moderate exercisers, improved well-being in all active Tx groups	109[d]
Male and female undergraduate students	80	35	active, aerobically fit vs. inactive subjects, one single 20-min bicycle exercise or waiting list	POMS; cardiovascular reactivity	improvement tension/anxiety in exercise groups; activity status increased vigor, no difference in cardiovascular reactivity	110[c]
Male and female VA employees; not depressed	70	mean 45	walk-jog, volleyball/weight lifting, waiting list 50–60 min 3 ×/week for 13 weeks	Multiple Adjective Checklist; Depression Adjective Checklist	significant increase in functional capacity, no effect on mood (dysphoria or euphoria)	111
Female college students	18	–	30 min aerobic exercise for 4 weeks exercise vs. no exercise condition	POMS; Multiple Affect Adjective Checklist	higher pulse rates and more vigor after exercise	16

POMS = Profile of Mood States; STAI = Spielberger Trait Anxiety Inventory; CES = Center for Epidemiological Studies.
[a] No control group.
[b] No random assignment.
[c] Improvement in psychological measures not dependent on physical fitness.
[d] Expectation measured.
[e] High drop-out rate.

concentration, feeling scared of the unknown, or worrying to occur more often in individuals who exercise, while individuals who seek out relaxation programs report more physiologic signs of anxiety, such as palpitations, sweating, nervousness, or headaches, suggesting that individuals who seek exercise may experience more cognitive symptoms than individuals who seek relaxation.

Table 1 shows that experimental studies have evaluated mostly male subjects. Studies that have included both men and women have, with one exception [10], found the psychological effects of exercise to be similar in women and men [11, 12]. Exercise of moderate intensity as indicated by sustained heavy breathing but not leading to exhaustion (of about 20–60 min duration, depending on individual tolerance, with a frequency of at least three sessions per week) reduces anxiety levels postexercise for 30 min to several hours even in the elderly [13]. By and large, increased cardiovascular fitness has been found associated with a greater sense of well-being and reduced anxiety. The early studies should be considered as merely suggestive and not conclusive because most lack the rigor of controlled clinical trials and most suffer from methodological and design problems. Few among the early studies have utilized random assignment [14, 15], observer ratings in addition to individual self-ratings, control groups [12], and standardized prolonged exercise practice, or have concurrently assessed cardiovascular fitness [16]. Nonetheless, most studies listed in table 1 demonstrate beneficial effects of exercise on temporary depressive feelings and confusion described by normal populations as well as increased energy levels and better concentration for up to one hour following aerobic exercise.

Effects of Exercise on the Personality

Physically active as opposed to sedentary individuals have been found to be more extroverted and less depressed [17]. Husman [18] examined the question whether a certain kind of personality may be prone to select different types of sports. On projective tests, boxers were found to be less aggressive than wrestlers while cross-country runners were more outwardly aggressive than boxers. The study indicated that the temperamental changes may well be state-dependent, since measurements of extra-directed and intra-directed aggressiveness changes with the training season. Another study by Morris and Husman [19] suggests that individuals whose lifestyles are less than satisfactory may be recruited into activities. Individuals in a conditioning class had consistently lower life quality scores than an inactive control group. Quality of life scores improved significantly with conditioning training. Another study, which compared meditators and runners on the Catell 16 Factor Personality Questionnaire, found runners to score lower on enthusiasm, assertiveness, and adventure than meditators or the population mean. Runners were more socially conscious and conscientious than meditators [20].

Studies that have examined the effects of conditioning programs on stable personality traits have given mixed results. These studies are hard to interpret due to the lack of suitable controls or nonrandom assignment (table 2). Unless positive expectations from exercise can be controlled, personality changes cannot with any certainty be attributed to exercise. Observations of greater self-assurance, self-sufficiency, sophistication, and more trusting tendencies reported on the Catell 16 Factor Personality Questionnaire after regular exercise for about 3 months could well be attributed to time, lifestyle, attention-support, or positive expectations. Ismail and Young [21] found no changes in personality factors, despite evidence of improved physical fitness. Buccola and Stone [22] compared 14 weeks of cycling and jogging in two groups of older men and found that despite similar improvements in aerobic capacity, only joggers changed significantly towards greater self-sufficiency and seriousness, albeit cyclers reported changes in the same direction. Jasnoski and Holmes [23], who studied undergraduates who participated in aerobic exercise for nearly 4 months, observed significant correlations between aerobic capacity and greater self-assurance and liberal-mindedness and reduced tension. In a controlled study with a smaller group of undergraduate women, Jasnoski et al. [24] and Jasnoski and Holmes [23] observed greater frustration tolerance with exercise training as opposed to no exercise but found no relationship between aerobic capacity changes and perception of abilities. The longest period of observations, 1 year, was reported by Jones and Weinhouse [25] in 12 male and female runners, who rated themselves as more intelligent, relaxed, assertive, and less serious after 1 year of regular running [26, 27].

A drawback to all personality research is the reliance on self-report in current standard personality inventories. Certain personality characteristics, especially maladaptive tendencies, regularly escape the individual's awareness and hence changes in these traits would not be recorded but might be more validly measured through observer ratings. Furthermore, data collection ought to be repeated over time during the course of exercise or a running program to detect variability and ought not to be limited to a

Table 2. The effect of conditioning training on personality factors

Subjects	Number	Age	Fitness	Experimental measures	Observed effect	Ref. No.
Normal men	53	17–59	10-week jogging program	Catell 16 PF, Taylor Manifest Anxiety Scale, physical fitness	aerobic capacity change correlated with more trusting and greater sophistication	26[a]
Faculty, staff members and local businessmen	58	21–61	90 min 3 ×/week mixed exercise for 4 months	Catell 16 PF, Eysenck Personal Inventory, Anxiety Scale, Biochemical Parameters	improved physical fitness but no change in personality; selected sample of 28 increases in conscientiousness	21
Male volunteers	36	60–79	14 weeks cycling vs. jogging	Catell 16 PF, estimated VO_2 maximum reaction time	improved aerobic capacity and reduced BP; joggers only became more self-sufficient and serious	22[b]
Women college students	103	mean 20	15 weeks aerobic training program, dancing, training, rope jumping 1 h 2 ×/week	Catell 16 PF, Zung Depression Scale, type A personality	aerobic capacity change correlated with reduced tension, greater self-assurance, more liberalness	23[a]
Women college students	39		2 ×/week 1 h training for 10 weeks, waiting list and independent controls	aerobic capacity, self-perception and confidence in abilities	exercise training group greater frustration tolerance not correlated to perception of abilities	24
Runners	12 7 M 5 F	23–41	1-year running program 45 min 3 ×/week	Catell 16 PF, cardiac output, personality inventory	more intelligent, relaxed, assertive, and feeling less serious	26[a]
Young and middle-aged male and female volunteers	32	28–30 50–53	1 h 3 ×/week for 10 weeks	heart rate, BP Life Satisfaction Control Scale, Depression and Anxiety Scales	reduction in anxiety improvd health status rating, young females increase in control	27

Catell 16 PF = Catell 16 Factor Personality Questionnaire. For further information, see footnotes to table 1.

cross-sectional before-and-after design. Keeping in mind that few of the studies have met acceptable standards of methodology, it might neverthe-less be concluded that there is tentative evidence for personality changes in the direction of greater confidence and less tension. These characteristics may or may not be related to the improved cardiovascular fitness which regularly accompanies long-term exercise.

Emotional Well-Being of Athletes (table 3)

Morgan's group [27, 28] have proposed that success in sports is inversely correlated with psychopathology and that the emotional balance of professional athletes is superior to that of the general population. In a series of studies, Morgan et al. [29] found that elite athletes were charac-terized by positive mental health. On the Eysenck Personality Inventory and the Profile of Mood States, athletes ranked below average on neuroti-cism, tension, depression, fatigue, and confusion compared to the popula-tion mean and showed significantly higher than normal scores on extrover-sion and vigor reflected in the so-called iceberg profile. Among wrestlers, rowers, and distance runners, the successfully competing athletes scored below unsuccessful athletes on negative feelings but scored similarly or higher on anger and on vigor and extroversion. Champion athletes on the American and British Track team scored significantly lower on anxiety than normals on the Catell 16 Personality Factor Scale while anxiety scores of junior athletes approximated the norm in a study by Heusner [30].

Euphoria and the Runner's High

No systematic research exists on the phenomenon of the runner's high, operationally defined as a 'euphoric sensation experienced during running, usually unexpected, in which the runner feels a heightened sense of well-being, enhanced appreciation of nature, and transcendence of barriers of time and space' [31]. In Callen's [32] sample of runners, three-fourths of whom were men and who ran at least 29 miles per week, two-thirds reported a runner's high sensation and it occurred in 44% of all runs. Thus, even though the runner's high is an unpredictable phenomenon both con-cerning the timing of the event and the percentage of runners experiencing euphoric sensations, individual accounts vividly describe the experience. Mandell's [33] exalted, yet articulate report describing powerful mystical and emotional sensations 'a cosmic view and peace are located between six and ten miles of running' remains the most quoted.

Table 3. Psychological characteristics of athletes

Subjects	Number	Age	Fitness	Experimental measures	Observed effect	Ref. No.
Athletes	54	20–30	basketball players, swimmers wrestlers, tennis players, cross-country runners	Eysenck Personality Inventory	cross-country runners more introverted than other athletes	29
Athletes	11	20–49	marathon runners	Eysenck Personality Inventory Anxiety Battery; Depression Adjective Checklist	within normal limits on EPI and depression, lower than normal on anxiety	27
Female athletes	27	–	1,500–3,000 m; 10,000 m; 42.2 km	24-hour history; STAI; Body Awareness Scale; POMS; Eysenck Personality Inventory	no differences in mood, personality traits or states between elite and regular runners POMS iceberg profile compared to average population	28
Male and female runners	23	mean 30	experienced runners vs. employees 30 min treadmill at 80% VO$_2$ max	POMS; plasma B-endorphin; corticotropin (ACTH); growth hormone (GH)	improved mood, tension, depression, anger, confusion immediately after the run; for males only, inverse association between mood, B-endorphin, and ACTH	94

POMS = Profile of Mood States; STAI = State Trait Anxiety Inventory.

A necessary prerequisite for the experience of a 'high' seems to be a relaxed attitude, a quiet unperturbed mind, and a total involvement in the movement with gentle running and no time commitment. The intensity of some of the sensual experiences, brighter colors, intense smells, clear sounds and the transcendental state bear similarities to the trance induced by hypnosis and to the manic state. Not surprisingly, it has been suggested [32, 34] that running and hypnosis increase right brain activation.

Gender Differences

Overall, few gender-dependent differences in the effects of exercise on mood have been reported. The fact that more women describe antidepressant effects may be related to the higher incidence of depressive symptoms in the adult female population [1]. More men tend to be serious runners and men tend to experience trance states more often than women who prefer imagery [32]. The strongest evidence supporting gender differences was reported by Kowal et al. [34] in a study of 200 male and female recruits and another 200 recruits after completion of basic training. Men improved in physical fitness, mood, self-concept, and attitude towards exercise, whereas female trainees showed no change. The major flaw in Kowal's study design was that different groups of men and women were tested before and after basic training. Not surprisingly, most studies [12, 34] have found men to have greater aerobic and work capacity than women.

Adverse Effects of Exercise

Biological organisms are characterized by an individually and constitutionally determined range of optimal functioning. Excessive exercise, in particular excessive running, can have harmful physical, psychological, and social consequences.

Staleness

Staleness is a term used for a pathological state of exhaustion and excessive weariness in response to an increasing training load or overtraining in athletes. The psychological and physiological changes bear resemblance to endogenous depression. Individuals report increased tension and depression, chronic fatigue, appetite loss, insomnia, decreased libido, and a reduction in functional capacity as well as metabolic and endocrine changes [34–37]. Overexertion and overcompetitiveness can increase tension and anxiety [38, 40]. Short-term high-intensity exercise increased tension and fatigue in female volunteers regardless of fitness status [38]. Morgan et al. [40–42] have shown the so-called inverted iceberg profile during overtraining and suggest that monitoring of the mood state using the POMS can be as informative as recording other stress markers such as heart rate, blood pressure, cortisol or catecholamine levels. Knapik et al. [43] reported greater than 25% decrements in performance and vigor and an 80% increase in fatigue in 89 soldiers following a maximal effort 20-km road march during which they carried a load of 46 kg.

Compulsive Exercise

Running has been said to be 'addictive' [35, 44]. This occurs when the commitment to the daily running routine takes on a higher priority than work, family, friends, or social functions and when running is continued despite serious bodily injuries. Another term used has been exercise dependence, which refers to the fact that many exercise fanatics experience acute, sometimes disabling, symptoms of anxiety, guilt and depression, when they are prevented from running [44, 45]. In the milder form the person requires daily exercise for optimal functioning, but can cope without daily running. In Callen's [32] sample, 25% said they had experienced depression, anger, frustration, with not being able to run due to an injury. So far, no studies have been done on persons who become 'addicted' to running. It is questionable, however, whether behavior that has become a compulsive habit should be defined as an addiction, since withdrawal symptoms alone do not define an addiction. Operationally, excessive exercise is maladaptive, and it might be a way to compensate for personal deficits. According to anecdotal reports, compulsive runners share a low self-esteem, self-dissatisfaction, a desire for perfectionism, and a decreased frustration tolerance. Touyz et al. [46] have described overexercisers whose initial pursuit of fitness eventually led to a pursuit of thinness and ultimately precipitated anorexia nervosa in both men and women. The clinical parallels that have been drawn between obligatory running and anorexia nervosa are mostly speculative and, so far, have not been supported by systematic data [44]. Excessive exercise in anorexia nervosa despite profound weight loss is well documented [47, 48]. A link between exercise and weight preoccupation in the normal population is suggested by the observation that male and female undergraduates reported hours of jogging correlated positively with abnormal eating attitudes and dieting [49].

Athlete's Neurosis

Little [50] coined the term athlete's neurosis on the basis of observations of men in their 30s and 40s who were referred for psychiatric treatment and who seemed excessively preoccupied with physical fitness. The first group were athletic men, whose emotional breakdown was initiated by a direct threat to their physical well-being, in 72.5% of the cases in the form of a physical injury or illness. Thus, these men became extremely sensitive to threats to their overvalued physical fitness. In nonathletic men, a direct physical threat had preceded symptoms in only 10.7%. Although all men reported symptoms of anxiety and a reactive depression, the latter group, called 'athletic neurotics' as opposed to the nonathletic neurotics, reported a happy and healthy premorbid personality, a low incidence of physical and psychiatric morbidity, good personal relationships and excellent physical health. By contrast, the life-long neurotics reported a high incidence of psychological and physical morbidity during childhood and in their families; they had been more introverted and less sociable, and their reaction was considered an expression of unresolved conflicts. Little [50, 51] considered the reaction of fitness fanatics a deprivation state, akin to a bereavement reaction to a loss of an overvalued aspect of their self-concept, their physical fitness and prowess. Athletic neurotics were found to be significantly more resistant to treatment than nonathletic neurotics. In this connection, a recent study of recreational runners is of interest in which a higher incidence of affective disorders was found in female but not in male runners compared to orthopedic patients [52].

The Effect of Exercise on Symptomatic Populations

Surprisingly little work on the effects of exercise in anxiety states or depressive illness has been done considering the beneficial mood elevating effects of exercise in healthy individuals. Published studies have mainly focussed on attenuation and relief from symptoms of depression. But, methodological problems have also plagued the research on the antidepressant effect of exercise in symptomatic populations [14]. Also, since anxiety and depression often coexist, although one or the other feeling state might predominate, the separate assessment of the two mood states is somewhat artificial. In depression a distinction needs to made between fleeting feelings of sadness, a temporary depressive mood state, lifelong unhappiness classified under dysthymia, or the major depressive disorder with more severe, unremitting, incapacitating symptoms.

The treatment of depressive symptoms through physical activity raises interesting theoretical and practical questions. Since exercise in part improves depressive symptoms through counteracting the postural and motor changes associated with depressive disorder, it may well be that the motor changes, such as retardation or agitation, are not merely epiphenomena, but specific to depression. Secondly, the antidepressant effect of exercise raises the possibility that peripheral activation, through behavioral changes, induces central effects. This notion was tested early in this century by Franz and Hamilton [53] who activated severely depressed patients by applying mechanical vibration along the spine. More recently, the idea was revived by Post and Goodwin [54] in their experiments on getting controls and patients to simulate certain components of the behavioral and clinical state of depression in order to assess whether alterations in amine metabolite levels were stress or activity related. Increases in physical activity over a 12-hour period significantly elevated urinary MHPG excretion in depressed patients, albeit without affecting the mood. In a further experiment, however, several moderately depressed patients reported feeling less depressed after simulating 'manic' behavior.

The studies on the effects of exercise in clinically depressed populations are summarized in table 4. The studies are largely based on two kinds of samples, men who had suffered a myocardial infarction and subsequently experienced a reactive depression and on mildly to moderately depressed subjects. The largest long-term, post-myocardial infarction sample was reported by Stern and Cleary [55] based on data from the National Exercise and Heart Disease Project. Despite a significant increase in work capacity in men who participated in regular exercise, no difference was found in psychosocial outcome compared to the no-exercise group. A possible explanation for the negative findings could be that the differences in exercise rates were marginal by 18 months. In the experimental group a mere 48% participated in at least half of the exercise sessions while 31% of the control group reported that they had begun to exercise shortly after randomization. Mayou [56] also failed to observe differences in either work capacity or psychological symptoms between exercising post-myocardial infarction patients and a standard treatment and advice group. All men had a comparatively good outcome and despite initial greater enthusiasm among exercisers, a fourth in each group eventually dropped out. In an uncontrolled study, Kavanagh et al. [57] found that MMPI depression

Table 4. The effects of exercise in symptomatic populations

Subjects	Number	Age	Fitness	Experimental measures	Observed effect	Ref. No.
Depressed, male postmyocardial infarction patients	44	n.s.	exercise rehabilitation slow distance jogging	MMPI depression scale; BP; aerobic capacity	exercise compliance associated with decreased depression after 4 years	57[a-c]
'Clinically' depressed students	100		jogging 3 or 5 days/week for 10 weeks	MMPI; Zung Self-Rating Depression Scale	depression, fatigue tension decreased; cheerfulness, energy increased	15
Male and female depressed outpatients	28	18–30	running 3–4 ×/week for 8 weeks vs. two kinds of individual psychotherapy	Hopkins Symptom Checklist-90	running as effective as individual psychotherapy	61
Men post-myocardial infarction	651	52	mixed physical activity 45 min 3 ×/week vs. no exercise prescription, 8 weeks to 18 months	Taylor Manifest Anxiety Scale, MMPI; psychosocial functioning	exercisers greater work capacity, but no greater psychosocial benefit	55
Male post-infarction patients	129	28–60	controls (n = 42); exercise (n = 44) 2 ×/week; advice (n = 44)	semistructured and standard psychiatric interviews	at 18 months similar work and psychological outcome	56[c, c]
Mildly depressed college women	43	n.s.	aerobic exercise 1 h 2 ×/week + other 15–20 min 4 ×/week; no exercise for 10 weeks	Beck Depression Inventory; expectancies physical fitness	aerobic exercise, greater improvement achieved first 5 weeks	67
Depressed male and female volunteers	74	30	running, meditation-relaxation, group therapy 1–2 h/week for 12 weeks	Hopkins Symptom Checklist; social adjustment; Cornell Medical Index	reduction in depression and anxiety for all groups at termination	62[e]

In-patients with major depression	43	40	aerobic exercise (n = 28), 1 h 3 ×/week up to 9 weeks occupational therapy (n = 21)	Beck Depression Inventory; Depression Analogue Scale; physical conditioning	aerobic exercise, greater improvement in depression scores	60
Delinquent male adolescents	69	14–18	40 min exercise 3 ×/week or baseball, volleyball instructions for 3 months	Piers Harris Children Self-Concept Scale; Beck Depression Inventory	improvement in mood, self-concept, and physical fitness in aerobic exercise greater than in comparison group	112[c]
Anxious volunteers	53	20–60	moderate exercise 20 min 4 ×/week for 10 weeks vs. attention flexibility training program	POMS; coping assets	exercisers greater reduction in tension, anxiety, confusion improvement in coping assets and well-being	39
Symptomatic women with mitral valve prolapse	32	21–49	aerobic exercise 3 ×/week 45 min for 12 weeks vs. no exercise group	STAI; General Well-Being Schedule, plasma catecholamine levels	improvement in well-being; reduction in State and Trait Anxiety; no differences in NE or E	113

POMS = Profile of Mood States; STAI = State Trait Anxiety Inventory; n.s. = not specified.

[a] No control group.
[b] No random assignment.
[c] Improvement in psychological measures not dependent on physical fitness.
[d] Expectation measure.
[e] High drop-out rate.

scores significantly improved over 4 years of rehabilitation, but they could not demonstrate an association between improved mood and cardiac fitness in men with coronary artery disease. In an earlier study [58], men who underwent endurance training were physiologically and psychologically indistinguishable from men who used autohypnosis. Only in those who exercised strenuously, fitness levels exceeded those of controls.

By contrast, studies in physically healthy moderately depressed subjects or explicitly recruited outpatients suggest that physical activity has clear antidepressant effects, even though the duration of these effects remains undetermined. In addition to an uncontrolled report in two severely depressed patients by Blue [59], Martinsen et al. [60] demonstrated a significant effect of aerobic exercise in more severely depressed inpatients who were randomly assigned and whose diagnosis was based on standardized criteria. However, the controls seem to have been more severely ill since twice as many, two-thirds, were on medication as opposed to one-third in the exercise group. Greist et al. [61] found in outpatients that the antidepressant effect of running equalled that of time-limited psychotherapy and was better than long-term psychotherapy. In another study [62] with random assignment, clinically depressed volunteers experienced similar reduction in depression and anxiety from either running, mediation-relaxation, and group therapy, the latter being the least effective. Table 4 indicates that mildly depressed or anxious volunteers consistently report a reduction in tension, anxiety, and depression, and an improved self-concept and well-being. One explanation for the scarcity of studies in depressive disorders may be the difficulty in activating seriously depressed patients. The lethargy, loss of interest, and feeling bad associated with the depressive state is the direct opposite of the motivation, hopefulness, energy, increased activity level associated with exercise.

Hypotheses Regarding the Mechanisms of Action Involved in Mood Alterations

Psychological Mechanisms

The expectation that exercise is beneficial for one's health, weight, and longevity influences subjective perception and thus may lead to feeling good through a sense of achievement and fulfillment. For instance, Leonardson and Garguilo [63] observed that subjects' perception of their fitness level correlated better with psychological measures than their actual fitness. It has also been proposed that exercise might improve mood through diversion from stressful stimuli; this has been called the distraction hypothesis [64]. For this reason, control groups with expectations for change which would be similar to those held by the exercising subjects or might be in similar ways distracted [65] would be indispensable, but few studies have controlled for expectation [66–68]. Similarly, the nonspecific effects of conducting exercise studies, such as increased attention and social interaction as a result of participation in the study, should be equalized across control groups. Conversely, negative attitudes about exercise capacity in disabled welders were closely related to the clinical grade of breathlessness [69].

Physiological Mechanisms

The reduction in symptoms of tension/anxiety which occurs with exercise of moderate and vigorous intensity may conceivably be mediated centrally through increased alpha rhythm and hemispheric synchronization [70]. Peripherally, local contralateral increases in cortical cerebral blood flow have been shown to occur during vigorous exercise by Olesen [71], making it likely that whole body exercise might increase blood flow in cortical areas. de Vries et al. [72–76] have suggested that the relaxing or 'tranquilizer' effect of exercise may be due in part to a decrease in resting muscle action potential (MAP) in the biceps and quadriceps muscles. This reduction in neuromuscular electrical hyperactivity was most pronounced in 'nervous' subjects. Investigators who have used the frontalis muscle have failed to find reductions in electrical activity [77]; however, as de Vries and co-workers [73, 74] have documented, the frontalis muscle might not be typical of the general skeletal musculature and not bear much relation to general muscle tension, but may relate to headaches. de Vries and Adams [75] compared the effect of 400 mg of meprobamate, exercise, and quiet rest in middle-aged to elderly men and women on muscular tension and found postexercise electrical activity significantly lower than during either drug or control condition. It has been suggested that the rise in body temperature with vigorous exercise [78] facilitates a relaxation response through a decrease in muscle spindle activity. Another explanation is that the random and intermittent proprioceptive stimulation with exercise, which would compete with excessive chronic muscle stimulation contributes to the reduction in muscular tension.

Research on the relationship between cardiovascular fitness and physiological responses has consistently demonstrated greater sympathetic arousal as measured by systolic blood pressure increases and skin temperature decreases [79], as well as reduced cardiovascular reactivity to stress with increasing fitness [80]. More fit subjects showed reduced heart rates, pre-ejection period, and systolic blood pressure responses to stressful behavioral tasks. Cox et al. [81] observed a positive correlation between aerobic power and the speed of recovery from stress in undergraduate students, while Dienstbier et al. [82] demonstrated reduced physiological responses to stressful sound and reduced capillary restriction in committed runners, although in their study no differences in galvanic skin response were found between runners and controls.

Biochemical and Endocrine Mechanisms

Monoamines. Exercise affects the sympathetic nervous system, including the adrenal medulla and the opioid-hypothalamic-pituitary-adrenal axis. Many hypotheses concerning the biochemical and endocrine regulation of mood have been advanced; in all likelihood, several interacting neurochemical systems regulate mood. The sympathetic nervous system receptors that have been found necessary to sustain intense and extended physical effort have also been implicated in influencing and improving emotional experiences and, especially, mood. Conversely, dysregulation of CNS aminergic receptors has been implicated in the pathogenesis of depression [83] and anxiety [84]. A sizable literature reviewed by Chaouloff [85] suggests that physical exercise in rats affects both brain norepinephrine and dopamine synthesis and metabolism, while exercise-induced lipolysis could increase plasma free tryptophan and as a result might lead to rises in brain serotonin. Studies in depressed or exercising humans have mostly sampled urinary or plasma monoamines and their metabolites. Depression has been found associated with a reduction in MHPG levels, a urinary metabolite of norepinephrine [83]. Epinephrine and norepinephrine secretion are clearly related to the relative intensity of the work in highly trained runners. Exercise-induced increases in circulating epinephrine levels were found to be greatly reduced by endurance training [86]. Since it is uncertain whether peripheral amine levels reflect central amine metabolism and since it is not known which areas of the brain might be involved in mood regulation, the aminergic hypothesis remains viable, but unproven. Since Post and Goodwin [54] have shown increases in CSF amine metabolites with increased motor activity, it cannot be ruled out that amine turnover changes in depression or exercise reflect changes in activity levels rather than in mood. Tai Chi movements [86, 87], which correspond to moderate exercise, have been found to increase heart rate and urinary norepinephrine secretion, and to decrease salivary cortisol levels. An unresolved biochemical puzzle is the significantly increased blood lactate levels associated with exercise in the presence of anxiety reduction in light of the fact that lactate infusions reliably induce panic attacks in predisposed individuals [84].

Opioids and the Hypothalamic-Pituitary-Adrenal Response
Currently, the most popular hypothesis involves changes in central opioid activity with aerobic exercise. Plasma beta-endorphin level elevations have been consistently demonstrated after running [88, 89], with greater increments in men than in women [90]. Marked increases in serum beta-endorphin up to 200 per unit have been reported after long-distance running [91]. It is important to realize that large interindividual variations in plasma beta-endorphin and leuenkephalin levels exist, and stress as a nonspecific stimulus cannot be ruled out. Thus, the increased cortisol plasma levels that have been reported in women runners and highly trained male runners [92] probably constitute an adaptation of the hypothalamic-pituitary-adrenal axis to the generalized stress of exercise. The assumption that endorphins might mediate an addictive dependence to running and generate euphoric feelings is attractive in light of the opioid theory of addiction, but, to date, no solid evidence has shown endorphin plasma levels or opioid receptor activity to be directly related to human addiction. Since beta-endorphins do not cross the blood-brain barrier, to date no study has demonstrated CNS increases in beta-endorphin levels which could be linked to the psychological effects. Among the pharmacologic probes, only naloxone, which is an effective antagonist for the mu receptor, the receptor that mediates analgesic action, has been tested. Markoff et al. [88] administered 0.8 mg naloxone s.c. to amateur runners in a placebo-controlled trial immediately following the run. Although significant post-run reductions in tension-anxiety and anger-hostility were reported, naloxone administration had no effects on any of the post-run POMS scores. By contrast, a study by Janal et al. [93] who administered two doses of 0.8 mg i.v. immediately and again at 20 min following the run to experienced long distance runners in a double-blind trial found naloxone to reverse post-ischemic pain and to attenuate rises in joy on euphoria ratings. The authors did not test whether the significant plasma level rises of beta-endorphin immunoreactivity, ACTH, GH, and PRL were influenced by naloxone administration. Kraemer et al. [94] observed signif-

icant correlations between negative mood and ACTH and beta-endorphin levels before exercise for males only, but not after treadmill exercise. Whether conditioning leads to neuroendocrine adaptations that would make the organism more stress resistant is undetermined, but preliminary evidence supports enduring homeostatic changes. Luger et al. [92] have demonstrated resting hypercortisolemia in highly trained runners, with an attenuated ACTH and cortisol secretory response to CRH, suggesting that conditioning may be associated with a reduction in pituitary-adrenal activation. Sinyor et al. [89] showed an earlier and larger initial plasma catecholamine and prolactin rise with mental stress in highly fit as opposed to low fit subjects and a subsequent more rapid return to baseline. In highly trained runners, Farrell et al. [89, 96] found increases in epinephrine, but no co-release of leuenkephalin from the adrenal medulla, suggesting that the enkephalin response to exercise may be dependent upon the training status of the subjects.

Conclusions

Mood

Regular exercise improves physical fitness and well-being across all ages and sexes. The findings are supported by psychosomatic, physiological, and neuromuscular evidence. Acute and long-term physical activity of a vigorous nature (exceeding 70% of maximum heart rate or age-adjusted maximum heart rate from 20 to 60 min with a frequency of at least 3 sessions per week) consistently reduces state anxiety and symptoms of depression for 30 min to several hours after exercise. Meditation may be comparable to exercise in its effectiveness to reduce anxiety levels, but meditation, yoga, or resting controls show no improvement in cardiovascular fitness, diastolic blood pressure, or bone mineral content.

Personality

Physically active and athletic individuals have been found to be more extroverted in their personality than sedentary individuals. Individuals whose lifestyles are less than satisfactory may be more prone to take up exercise. The data on the relationship between personality types and different kinds of sports are inconclusive. Personality features of athletes show seasonal changes.

With prolonged exercise, runners seem to become more self-sufficient, serious, and more relaxed than before running; cycling produces trends in the same direction.

Athletes

Success in sports seems to be inversely related to psychopathology. Elite athletes report a higher level of well-being and better emotional adjustment than the general population.

Athletes as a group score within the normal range on energy, extroversion, and on negative feeling states.

Runner's High

The runner's high is a kind of euphoria in which the runner feels a heightened sense of well-being and enhanced sensual experiences. The sensation occurs unexpectedly and cannot be predicted. It seems to correspond to an altered state of consciousness, resembling a trance or the manic state.

Adverse Effects of Exercise

Overexercising can have harmful physical (staleness), psychological (depression), and social (social isolation) consequences.

Exercise becomes maladaptive if it becomes a compulsive habit. This occurs if individuals develop disabling symptoms of guilt, depression, or anxiety when they are prevented from running.

The athlete's neurosis is an emotional breakdown initiated by a threat to the individual's self-perception in the form of a physical injury or illness. Psychologically, the reaction bears similarity to bereavement and represents a reaction to a loss of an overvalued aspect of the self, namely physical fitness and prowess. Although premorbid adjustment has been found to be excellent, the prognosis is guarded.

Depression and Anxiety

Studies in depressive disorders suggest a powerful therapeutic effect of regular exercise. Beneficial effects are less obvious in reactive depressions following a myocardial infarction. Better controlled and more systematic investigations in populations suffering from depressive disorders are necessary to provide conclusive evidence. Studies on the effects of exercise in anxiety disorders are lacking.

Hypotheses about the Mechanisms of Action Involved in
Mood Alterations

Many hypotheses concerning the mechanisms leading to mood improvement have been tested more or less independently. In all likelihood, a host of factors interact to produce the beneficial effects.

The psychological influences include expectation, distraction, self-motivation, social interaction, and therapeutic attention.

Physiologically improved cardiovascular fitness, increased cortical cerebral blood flow, hemispheric synchronization and reduced resting muscle action potential have been documented.

Biochemically, the response to exercise resembles the stress response with increased peripheral and central catecholamine levels, activation of the opioid-hypothalamic-pituitary-adrenal axis with co-release of beta-endorphins, but not necessarily leu-enkephalin.

Finally, the studies offer compelling evidence that increased movements and regular aerobic exercise promote health and coping with stress. Future research ought to address the role of exercise in disease prevention and treatment, in particular for anxiety and depressive disorders and stress-related physical disorders and, furthermore, investigate more systematically which psychological or biological mechanisms contribute to the beneficial effects of exercise.

References

1   Weissman MM, Klerman GL: Sex differences and the epidemiology of depression. Arch Gen Psychiatry 1977;34:98–111.
2   Spielberger CD: Manual for the State-Trait Anxiety Inventory, revised ed. Palo Alto, Consulting Psychologists Press, 1983.
3   Hathaway SR, McKinley JC: The Minnesota Multiphasic Personality Inventory. Minneapolis, University of Minnesota Press, 1943.
4   Taylor JA: A personality scale of manifest anxiety. J Abnorm Soc Psychol 1953;48: 285–290.
5   McNair D, Lorr M, Droppleman L: Manual: Profile of Mood States. San Diego, Educational & Industrial Testing Service, 1971.
6   Beck AT: Depression Inventory. Philadelphia. Center for Cognitive Therapy, 1978.
7   Catell R: The Sixteen Personality Factor Questionnaire. Champaign, Institute for Personality and Ability Testing, 1972.
8   Eysenck HJ, Eysenck GBG: Manual of the Eysenck Personality Inventory. London, University of London Press, 1964.
9   Schwartz GE, Davidson RJ, Goleman DJ: Patterning of cognitive and somatic pro-

cesses in the self-regulation of anxiety: Effects of meditation versus exercise. Psychosom Med 1978;40:321–328.

10   Wood D: The relationship between state anxiety and acute physical activity. Am Correct Ther J 1977;13:110–117.

11   Young RJ: The effect of regular exercise on cognitive functioning and personality. Br J Sports Med 1979;13:110–117.

12   Wilson VE, Berger BG, Bird EI: Mood profiles of marathon runners, joggers, and non-exercisers. Percept Mot Skills 1981;53:472–474.

13   Blumenthal JA, Emery CF, Madden DJ, et al: Cardiovascular and behavioral effects of aerobic exercise training in healthy older men and women. J Gerontol 1989;44: M147–157.

14   Berger B, Owen DR: Mood alteration with swimming – swimmers really do 'feel better'. Psychosom Med 1983;45:425–433.

15   Brown RS, Ramirez DE, Taub JM: The prescription of exercise for depression. Phys Sports Med 1978;6:34–49.

16   Flory JD, Holmes DS: Effects of an acute bout of aerobic exercise on cardiovascular and subjective responses during subsequent cognitive work. J Psychosom Res 1991; 35:225–230.

17   Eysenck HJ, Nias DKB, Cox DN: Sport and personality. Adv Behav Res Ther 1982; 4:1–55.

18   Husman BF: Aggression in boxers and wrestlers as measured by projective techniques. Res Q 1955;26:421–425.

19   Morris AF, Husman BF: Life quality changes following and endurance conditioning program. Am Corr Ther J 1978;32:3–6.

20   Large E, Hartung G, Borland C: Runners and meditators: A comparison of personality profiles. J Pers Assess 1979;43:501–502.

21   Ismail AH, Young RJ: Effect of chronic exercise on the personality. Ann NY Acad Sci 1977;301:958–969.

22   Buccola VA, Stone WJ: Effects of jogging and cycling programs on physiological and personality variables in aged men. Res Q 1975;46:134–139.

23   Jasnoski M, Holmes D: Influence of initial aerobic fitness, aerobic training and changes in aerobic fitness on personality functioning. J Psychosom Res 1981;25: 553–556.

24   Jasnoski M, Holmes D, Solomon S, Agular C: Exercise, changes in aerobic capacity, and changes in self-perceptions: An experimental investigation. J Res Personal 1981;15:460–466.

25   Jones RD, Weinhouse S: Running as therapy. J Sports Med 1979;19:397–404.

26   Hammer WM, Wilmore JH: An exploratory investigation in personality measures and physiological alterations during a 10-week jogging program. J Sports Med Phys Fitness 1973;13:238–247.

27   Morgan WP, Costill DL: Psychological characteristics of the marathon runner. J Sports Med 1972;12:42–46.

28   Morgan WP, O'Connor PJ, Sparling PB, et al: Psychologic characterization of the elite female distance runner. Int J Sports Med 1987;8:124–131.

29   Morgan WP: Selected physiological and psychomotor correlates of depression in psychiatric patients. Res Q 1968;39:1037–1043.

30 Heusner L: Personality traits of champion and former champion athletes. Research Study. Champaign, University of Illinois, 1952.

31 Sachs ML: The runner's high; in Sachs ML, Buffone GW (eds): Running as Therapy: An Integrated Approach. Lincoln, University of Nebraska Press, 1984.

32 Callen KE: Mental and emotional aspects of long-distance running. Psychosomatics 1983;24:133–151.

33 Mandell AJ: The second second wind. Psychiatr Ann 1979;9:57–69.

34 Kowal D, Patton J, Vogel J: Psychological states and aerobic fitness of male and female recruits before and after basic training. Aviat Space Environ Med 1978;49: 603–606.

35 Morgan WP: Negative addiction in runners. Physician Sports Med 1979;7:57–70.

36 Morgan WP: Coping with mental stress: The potential limits of exercise intervention (final report). Bethesda, NIMH, 1984, pp 11–14.

37 Morgan WP, Brown DR, Raglin PJ, et al: Psychological monitoring of overtraining and staleness. Br J Sports Med 1987;21:107–114.

38 Steptoe A, Bolton J: The short term influence of high and low intensity physical exercise on mood. Psychol Health 1988;2:91–96.

39 Steptoe A, Edwards A, Moses J, et al: The effects of exercise training on mood and perceived coping ability in anxious adults from the general population. J Psychosom Res 1989;33:537–547.

40 Morgan WP, Costill DL, Flynn MG, et al: Mood disturbance following increased training in swimmers. Med Sci Sports Exerc 1988;20:408–414.

41 Morgan WP, Goldston SE (eds): Exercise and Mental Health. Washington, Hemisphere, 1987.

42 Morgan WP, Roberts JA, Brand FR, et al: Psychological effect of chronic physical activity. Med Sci Sports 1970;2:213–217.

43 Knapik CPTJ, Staab J, Bahrke M, et al: Soldier performance and mood states following a strenuous road march. Milit Med 1991;156:197–200.

44 Yates A: Compulsive Exercise and the Eating Disorders: Toward an Integrated Theory of Activity. New York, Brunner/Mazel Publishers, 1991.

45 Thaxton L: Physiological and psychological effects of short-term exercise addiction on habitual runners. J Sport Psychol 1982;4:73–80.

46 Touyz SW, Beumont PJV, Hook S: Exercise anorexia: A new dimension in anorexia nervosa?; in Beumont PJV, Burrows GD, Casper RC (eds): Handbook of Eating Disorders. I. Anorexia and Bulimia Nervosa. Amsterdam, Elsevier, 1987, pp 143–157.

47 Kron L, Katz JL, Gregory G, et al: Hyperactivity in anorexia nervosa: A fundamental clinical feature. Compr Psychiatry 1978;19:433–440.

48 Casper RC, Schoeller DA, Kushner R, et al: Total daily energy expenditure and activity level in anorexia nervosa. Am J Clin Nutr 1991;53:1143–1150.

49 Richert AJ, Hummers JA: Patterns of physical activity in college students at possible risk for eating disorder. Int J Eat Dis 1986;5:757–763.

50 Little JC: The athlete's neurosis: A deprivation crisis. Acta Psychiatr Scand 1969;45: 187–197.

51 Little JC: Neurotic illness in fitness fanatics. Psychiatr Ann 1979;9:49–56.

52 Colt EWD, Dunner DL, Hall K, Fieve RR: A high prevalence of affective disorders in runners; in Sacks MH (ed): The Psychology of Running. Illinois, Human Kinetics Publishers, 1981, pp 234–238.

53 Franz SI, Hamilton GV: The effects of exercise upon the retardation in conditions of depression. Am J Insanity 1905;62:239.

54 Post RM, Goodwin FK: Simulated behavior states. Biol Psychiatry 1973;7:237–254.

55 Stern MJ, Cleary J: The National Exercise and Heart Disease Project. Arch Intern Med 1982;142:1093–1097.

56 Mayou A: A controlled trial of early rehabilitation after myocardial infarction. J Cardiac Rehabil 1983;3:397–402.

57 Kavanagh T, Shepherd RJ, Tuck JA, et al: Depression following myocardial infarction: The effect of distance running. Ann NY Acad Sci 1977;301:1029–1038.

58 Kavanagh T, Shepherd RJ, Pandit V, et al: Exercise and hypnotherapy in the rehabilitation of the coronary patient. Arch Phys Med Rehabil 1970;51:578–587.

59 Blue RF: Aerobic running as a treatment for moderate depression. Percept Mot Skills 1979;48:228.

60 Martinsen EW, Medhus A, Sandvik L: Effects of exercise on depression. A controlled study. Br Med J 1985;291:109–112.

61 Greist JH, Klein MH, Eischens RR, et al: Running as treatment for depression. Compr Psychiatry 1979;20:41–54.

62 Klein MH, et al: A comparative outcome study of group psychotherapy versus exercise treatment for depression. Int J Ment Health 1985;13:148–177.

63 Leonardson GR, Garguilo RM: Self-perception and physical fitness. Percept Mot Skills 1978;46:338–343.

64 Bahrke MS, Morgan WP: Anxiety reduction following exercise and meditation. Cog Ther Res 1978;2:323–333.

65 Fillingham RB, Roth DL, Haley WE: The effects of distraction on the perception of exercise-induced symptoms. J Psychosom Res 1989;33:241–248.

66 Dienstbier R: The effect of exercise on personality; in Sachs M, Buffone G (eds): Running as Therapy: An Integrated Approach. Lincoln, University of Nebraska Press, 1984.

67 McCann IL, Holmes DS: Influence of aerobic exercise on depression. J Pers Soc Psychol 1984;46:1142–1147.

68 Simons CW, Birkimer JC: An exploration of factors predicting the effects of aerobic conditioning on mood state. J Psychosom Res 1988;32:63–75.

69 King B, Cotes JE: Relation of lung function and exercise capacity to mood and attitudes to health. Thorax 1989;44:402–409.

70 Dishman RK: Medical psychology in exercise and sport. Med Clin North Am 1985;69:123–143.

71 Oleson J: Contralateral focal increase of cerebral blood flow in man during arm work. Brain 1971;94:635–646.

72 de Vries H: Immediate and long term effects of exercise upon resting muscle action potential level. J Sports Med Phys Fitness 1968;8:1–11.

73 de Vries HA: Physiology of Exercise. Dubuque, Brown, 1974.

74 de Vries HA, Beckman P, Huber H, et al: Electromyographic evaluation of the effects of sauna on the neuromuscular system. J Sports Med 1968;8:61–69.

75 de Vries H, Adams G: Electromyographic comparison of single doses of exercise and meprobamate as to effects on muscular relaxation. Am J Phys Med 1972;51:130–141.

76   de Vries HA, Wiswell RA, Bulbulian R, et al: Tranquillizer effect of exercise. Am J
      Phys Med 1981;60:57–66.
77   Farmer PK, Olewine DA, Comer DW, et al: Frontalis muscle tension and occipital
      alpha production in young males with coronary prone (type A) and coronary resis-
      tant (type B) behavior patterns: Effects of exercise. Med Sci Sports 1978;10:51–
      58.
78   Cannon JG, Kluger MJ: Endogenous pyrogen activity in human plasma after exer-
      cise. Science 1983;220:617–619.
79   Cantor JR, Zillman D, Day KD: Relationship between cardiorespiratory fitness and
      physiological response to films. Percept Mot Skills 1978;46:1123–1130.
80   Shulhan D, Scher H, Furedy JJ: Phasic cardiac reactivity to psychological stress as a
      function of aerobic fitness level. Psychophysiology 1986;23:562–566.
81   Cox JP, Evans JF, Jamieson JL: Aerobic power and tonic heart rate responses to
      psychosocial stressors. Pers Soc Psychol Bull 1979;5:160–163.
82   Dienstbier RA, Crabbe J, Johnson GO, et al: Exercise and stress tolerance; in Sachs
      MH, Sachs ML (eds): Psychology of Running. Champaign, Human Kinetics Publish-
      ers, 1981.
83   Schildkraut J: The catecholamine hypothesis of affective disorders: A review of
      supporting evidence. Am J Psychiatry 1965;122:509–522.
84   Charney DS, Heninger GR: Serotonin function in panic disorders: The effect of
      intravenous tryptophan in healthy subjects and panic disorder patients before and
      during alprazolam treatment. Arch Gen Psychiatry 1986;43:1059–1065.
85   Chaouloff F: Physical exercise and brain monoamines: A review. Acta Physiol Scand
      1989;137:1–13.
86   Koh TC: Tai Chi Chuan. Am J Chin Med 1981;9:15–22.
87   Li T: Effectiveness of Tai Chi Chuan in the treatment of hypertension; in: Exercise
      Therapy in Cardiovascular Diseases. Beijing, People's Sports Press, 1979.
88   Markoff R, Ryan P, Young T: Endorphins and mood changes in long-distance run-
      ning. Med Sci Sports Exerc 1982;14:11–15.
89   Farrell PA, Gustafson AB, Morgan WP, et al: Enkephalins, catecholamines, and
      psychological mood alterations: Effects of prolonged exercise. Med Sci Sports Exerc
      1987;19:347–353.
90   Gambert SR, Hagen TG, Garthwaite EH, et al: Exercise and the endogenous opiods:
      Letter to the editor. N Engl J Med 1981;305:1590–1591.
91   Appenzeller O, Standefer J, Appenzeller J, et al: Neurology of endurance training. V.
      Endorphins. Neurology 1980;30:418–419.
92   Luger A, Deuster PA, Kyle SB, et al: Acute hypothalamic-pituitary-adrenal re-
      sponses to the stress of treadmill exercise: Physiologic adaptations to physical train-
      ing. N Engl J Med 1987;316:1309–1315.
93   Janal MN, Colt EWD, Crawford Clark W, Glusman M: Pain sensitivity, mood and
      plasma endorphin levels in man following long-distance running: Effects of nalox-
      one. Pain 1984;19:13–25.
94   Kraemer RR, Dzewaltowski DA, Blair MS, et al: Mood alteration from treadmill
      running and its relationship to beta-endorphin, corticotropin, and growth hormone.
      J Sports Med Phys Fitness 1990;30:241–246.
95   Sinyor D, Schwartz SG, Peronnet F, et al: Aerobic fitness level and reactivity to
      psychosocial stress: Physiological, biochemical, and subjective measures. Psycho-
      som Med 1983;45:205–217.

96 Farrell PA, Gates WK, Maksud MG, et al: Increases in plasma beta-endorphin/beta-lipotropin immunoreactivity after treadmill running in humans. J Appl Physiol 1982;52:1245–1249.

97 Carr DB, Bullen BA, Skrinar GS, et al: Physical conditioning facilitates the exercise-induced secretion of beta-endorphins and beta-lipotropin in women. N Engl J Med 1981;305:560–562.

98 Popejoy DI: The effects of a physical fitness program on selected psychological and physiological measures of anxiety; Diss, University of Illinois, 1967. Diss Abstr 1968;29:4900A – University Microfilms No 68-08196.

99 Morgan WP, Roberts JA, Brand FR, Deinerman AD: Psychological effect of chronic physical activity. Med Sci Sports 1970;2:213–217.

100 Folkins CH: Effects of physical training on mood. J Clin Psychol 1976;32:385–388.

101 Blumenthal JA, Williams RS, Needels TL, et al: Psychological changes accompany aerobic exercise in healthy middle-aged adults. Psychosom Med 1982;44:529–536.

102 Lichtman S, Poser EG: The effects of exercise on mood and cognitive functioning. J Psychosom Res 1983;27:43–52.

103 Goldwater BC, Collins ML: Psychological effects of cardiovascular conditioning: A controlled experiment. Psychosom Med 1985;47:174–181.

104 Holmes DS, Roth DL: Association of aerobic fitness with pulse rate and subjective responses to psychological stress. Psychophysiology 1985;22:525–529.

105 Roth DL, Holmes DS: Influence of physical fitness in determining the impact of stressful life events on physical and psychological health. Psychosom Med 1985;47:164–173.

106 Hughes JR, Casal DC, Leon AS: Psychological effects of exercise: A randomized cross-over trial. J Psychosom Res 1986;30:355–360.

107 Roth DL, Holmes DS: Influence of aerobic exercise training and relaxation training on physical and psychological health following stressful life events. Psychosom Med 1987;49:355–365.

108 Jin P: Changes in heart rate, noradrenaline, cortisol and mood during tai chi. J Psychosom Res 1989;33:197–206.

109 Moses J, Steptoe A, Mathews A, et al: The effects of exercise training on mental well-being in the normal population: A controlled trial. J Psychosom Res 1989;33:47–61.

110 Roth DL: Acute emotional and psychophysiological effects of aerobic exercise. Psychophysiology 1989;26:593–602.

111 Lennox SS, Bedell JR, Stone AA: The effect of exercise on normal mood. J Psychosom Res 1990;34:629–636.

112 MacMahon JR, Gross RT: Physical and psychological effects of aerobic exercise in delinquent adolescent males. Am J Dis Child 1988;142:1361–1366.

113 Scordo KA: Effects of aerobic exercise training on symptomatic women with mitral valve prolapse. Am J Cardiol 1991;67:863–868.

Prof. Regina C. Casper, MD, Department of Psychiatry and Behavioral Sciences,
Room TD 114, Stanford University School of Medicine,
Stanford, CA 94305-5490 (USA)

Simopoulos AP, Pavlou KN (eds): Nutrition and Fitness for Athletes.
World Rev Nutr Diet. Basel, Karger, 1993, vol 71, pp 144–153

Vitamin Requirements for Increased Physical Activity: Vitamin E

Irene Simon-Schnass

Hermes Arzneimittel GmbH, Grosshesselohe, BRD

Introduction

With a few exceptions, the main reason for taking part in sports is to achieve optimal athletic performance. Naturally, the demands which the individual makes on himself vary considerably. The common element, however, is that this goal cannot be achieved without more or less intensive training. Looking at performance from a physiological point of view, training should generally be regarded as a constant adaptation to stress. The stimuli from a set training program impair homeostasis (exertion phase), in which catabolic processes are of major importance. Today, impairment of homeostasis is considered to be the cause of adaptive changes in the stressed system. The adaptation reactions cause homeostasis to be reactivated on a different level (regeneration phase). This is where anabolic processes dominate.

Physiological Background

Energy Metabolism

From an energy standpoint, each stress stimulus primarily attacks the cell. Each movement is associated with the turnover of substrates and thus with energy consumption. From a simplified point of view, blood circulation is no more than an aid to substrate delivery and removal. Circulation increases during dynamic muscular activity, as the increased consumption of substrate and the removal of end products would otherwise not be possible. However, muscular contraction also leads to vascular compression which in turn causes a regional and short-term reduction in circulation with limited hypoxia. This applies particularly to sports with concentrated power development. On the other hand, in sports with an emphasis on endurance and lower locally active forces, the transport capacity of the blood and oxygen exchange into the cell becomes the limiting factors. In both cases, transient oxygen deficiency can occur, despite the high oxygen turnover.

Cellular Structures

Classical sports physiology has concerned itself intensively with the phenomena of availability, turnover and regeneration of substrates for obtaining energy. In addition to this, a fact which has as yet hardly been acknowledged is that the effects on the structured components of the cell are of special relevance. The cell membrane contains important switching points for transport processes, as well as for reactive processes. It is no coincidence that the compartmentalization of the cell is a major feature of living structures. Indeed, this is the basis on which the cell is able to function. Membranes participate in some form or other in the vast majority of metabolic processes. The main objective, even in the case of intense physical activity which, as described above, leads to impaired homeostasis, is always to maintain the integrity of the membrane structures, or at least that changes be reversible.

Undesirable Effects of Energy Metabolism

The major cause of membrane damage is the formation of free radicals which can arise from various processes during metabolism. In the aerobic energy supply, most of the ATP is formed during endoxidation. This is when electrons of a substrate (e.g. pyruvate or succinate) are transferred via a redox chain to oxygen, the end product formed being water (reduction of the oxygen to water):

$$O_2 + 4H^+ + 4e \rightarrow 2\ H_2O\ \text{(water)}.$$

It is known that free radicals can arise in the case of incomplete reduction of oxygen. If less than 4 electrons are made available, the following activated oxyged species are formed:

$$O_2 + 1e \rightarrow {}^\cdot O_2^- \text{ (superoxide radical)},$$
$$O_2 + 2e + 2H^+ \rightarrow H_2O_2 \text{ (hydrogen peroxide)},$$
$$O_2 + 3e + 3H^+ \rightarrow H_2O_2 + {}^\cdot OH \text{ (hydroxyl radical)}.$$

At rest, 250–300 ml of oxygen per minute are usually taken up. During exercise, oxygen uptake can increase to 4,700 ml/min or even more, depending on training conditions. Around 3–10% of the metabolized oxygen is not completely reduced to water, but forms these different radicals.

However, other metabolic processes also lead to the formation of free radicals. Physically strenuous activity induces certain inflammatory reactions which are associated with the increased formation of radicals (e.g. leukocyte activation with phagocytosis, leukotriene synthesis). In addition, radicals from outside are taken up in the body, e.g. from air pollution (elevated ozone content in the air we breathe, car exhaust fumes), high ultraviolet (UV) radiation, cigarette smoke. This can also be of great relevance to athletes because the damaging reactive mechanisms of these radicals do not differ greatly from the reactions mentioned here.

One characteristic of free radicals is their high, in part extremely high reactive capability. The above-mentioned activated oxygen forms react very readily with other substances and thus form other radicals. Particular consideration should be given here to the reaction with lipids (formation of fatty acid radicals) and with proteins, particularly to those which contain functional SH groups (inactivation, formation of carbonyls). Destructive chain reactions can be set off which, if they are not interrupted by processes mentioned later, can lead to functional impairments up to complete destruction of the cell.

In the case of extreme physical activity (the term 'extreme' should always be considered of an individual nature), the following factors can lead to an increased discharge of free radicals. The oxygen turnover can be increased up to 20 times resting consumption. Correspondingly, the incomplete reduction of oxygen to water and the associated formation of free radicals can also increase. Hypoxic cells are particularly susceptible to oxidative stress. If there is not enough oxygen available to accept electrons, the electrons will be transferred to other low molecular weight molecules which in turn induce radical chain reactions. At low pH (metabolic acidosis during physical activity), $^{\bullet}O_2$ can be converted into the highly toxic $^{\bullet}OOH$ radical. $^{\bullet}O_2$ is derived from the respiratory chain whereas the required hydrogen is supplied by the lactic acid which is formed. In the case of physiological pH, only around 1% of the $^{\bullet}O_2$ is converted to an $^{\bullet}OOH$ radical. However, the percentage increases with increasing acidosis.

Adaptation

It has long been known that the organism responds adaptively to training stimuli. This adaptation consists, for example, of increased synthesis of the enzymes involved in energy supply, an increased storage of glycogen, the build-up of higher alkali reserves, a growth of muscle mass and an increase in the capillarization of the trained muscles as well as a hormonal adaptation. Moreover, all aerobic cells are forced to develop and maintain a defense system against oxidative damage. Enzymes as well as antioxidants are components of this system. The 'antioxidative strategy' of aerobic cells is targeted at inhibiting or blocking potentially toxic oxygen species or their derivatives at the various levels of formation or their reaction with biomolecules.

Enzymes. Typical 'detoxification enzymes' are superoxide dismutase (SOD), the catalases and various peroxidases. If the organism is flooded with radicals, as in case of athletic activity, these enzymes are reactively synthesized to an elevated degree. Naturally, a certain time-lag between the occurrence of the noxae (radicals) and the higher enzymatic level as a protective measure cannot be avoided. If the stimulus for the increased enzymatic formation does not persist, it drops back to its original level. This suggests that protection against enzymatic oxidation can only be effectively boosted by regularly taking part in sports. From this point of view, athletes who train irregularly and at varying degrees of intensity are relatively poorly protected against oxidative stress.

Antioxidants. On the other hand, the defence system comprises various antioxidants in addition to the antioxidative enzymes. Many substances that also occur physiologically represent antioxidants in vitro. To date, practical relevance has only been found for vitamins E and C as well as glutathione and, in certain cases, uric acid, taurine and the amino acids, cysteine and histidine as well as several xenobiotics. We will not go into these here.

Vitamin E is a lipophilic radical inhibitor. Correspondingly, its action is limited mainly to the region of the lipophilic membrane. Because it is located there in the immediate vicinity of the substances in danger of oxidation, vitamin E can very effectively circumvent the peroxidation of fatty acids as well as the oxidation of cholesterol and proteins. This is of great importance for membrane integrity as well as for preventing damage to the tissues, e.g. vascular intima.

During vitamin E deficiency, no other antioxidant could yet be proven to break the radical chain reaction in the lipophilic area of the cell. This means that vitamin E cannot be replaced by any other substances at its functional sites. In addition to the inhibition of lipid peroxidation in the biological membranes, vitamin E also appears to play a role in the repair of oxidized amino acids. In this way, vitamin E could also be attributed a special role in the repair of damaged amino acids in the integral membrane proteins [1–5].

Physical Activity and Vitamin E

Effect of Physical Activity on Vitamin E Status
As shown by animal experiments, endurance training caused greater reduction in the vitamin E content of liver and muscles [6, 7] compared to untrained animals. The vitamin E concentration in muscle was significantly reduced compared to the control animals. Other experiments showed that the increase in the number of mitochondria, an adaptation reaction to the increased energy consumption, was not accompanied by a parallel increase in vitamin E content. These experiments suggest that endurance training leads to increased consumption of vitamin E in the affected tissues and thus to elevated demand [8–11].

Effect of Vitamin E Deficiency on Performance
Reduced cellular respiration is regularly found in the mitochondria of vitamin E deficient experimental animals [12–14]. Various investigations have shown that the fundamental blockage of electron flow in the case of vitamin E deficiency already takes place at the beginning of the respiratory chain during the shift of the hydrogen of $NADH_2$ to an iron-sulphur protein (complex I) or during the conversion of the hydrogen from succinate to another iron-sulphur protein (complex II). Furthermore, the introduction of hydrogen into the respiratory chain is hindered by a reduction in the activity of lipoic acid amide dehydrogenase (LADH); LADH catalyses the conversion of the hydrogen of the lipoic acid to NAD and thus creates an important connective pathway between the final breakdown of the carbon skeleton (pyruvate) and the respiratory chain [15].

Interestingly, these are all vitamin E-sensitive proteins with unstable sulfhydryl groups. Conversion of these SH groups into the oxidized state leads to the loss of biological activity. It has been proven that vitamin E is localized in the immediate vicinity of these enzymes containing SH groups and has a protective effect on these SH groups. In this way, vitamin E plays an indirect, but major role in the electron transport of the respiratory chain and thus in the production of energy.

Therefore an action of vitamin E in the context of cellular function can be expected if specific purposes have to be fulfilled with the aid of aerobic energy supply [16, 17]. The significance of viamin E in aerobic energy supply was proven long ago. The fact that ATP synthesis is reduced during vitamin E deficiency was shown almost 40 years ago [18, 19]. This was confirmed in 1960 [20] and twice in 1971 [12, 14]. The ineffective phosphorylation is expressed by a reduced synthesis of ATP per molecule of oxygen, in other words, by a reduced P/O quotient [5, 13]. In addition to the impairments described, intramuscular creatine-kinase activity proved to be reduced, a condition which can lead to a retarded regenerative ability [21].

The major conclusion that can be drawn from the facts presented is that a positive effect of vitamin E on performance primarily occurs if aerobic energy production represents a major contribution to performance. Equally, measurable changes can only be expected if oxygen-related performance parameters (e.g. the determination of anaerobic threshold) are tested. In contrast, experimental designs which primarily include performance in the anaerobic range or sport-motoric indices as test parameters appear unsuitable for providing a valid assessment of the effect of vitamin E on energy supply and performance [22–25].

Effect of Vitamin E on Performance without Proven Vitamin E Deficiency

As far back as 1955, Cureton [26], and later Clausen [27] and Nagawa et al. [28], proved that the administration of vitamin E leads to improved ergometric performance. The results of Kobayashi [29] and our own investigations [30] yielded an improvement in performance during exertion at altitude with changes in the oxygen debt and elevated oxygen uptake at the anaerobic threshold.

Effect of Vitamin E Status on Regeneration and Prevention of Damage to Health

Using the regeneration quotient, the ratio of oxygen consumption during the exertion phase to oxygen consumption after exertion, it could be proven that the average regeneration phase was shortened by 10% on vitamin E supplementation [31].

In investigations on rats which were either fed with a vitamin E-rich diet or control feed and were exercised to exhaustion, it was found that the various organs of the control animals had elevated concentrations of carbonyl compounds. These carbonyl compounds arise from oxidative changes to proteins and can be seen as indicators of cellular damage [32]. In two experiments in man, the effect of vitamin E supplementation on the expiration of pentane, a marker of oxidative stress or increased lipid peroxidation, was investigated. In both cases, exhaustive physical exertion led to an increase in pentane production. Supplementation with vitamin E was able to significantly throttle this process [33, 34]. Our own investigations on high-altitude mountain climbers showed that the increase in pentane expiration found in the control group could be prevented by vitamin E supplementation [30]. These findings were confirmed by recent investigations on test subjects. The ones who had not received vitamin E supplementation showed oxidative membrane damage with enzymatic leakage into the plasma after exhaustive exercise [35].

Our own investigations showed that increased lipid peroxidation can moreover cause deterioration in blood-flow properties [36]. In these experiments, the vitamin E status was not obtained, but because we were dealing with healthy test subjects who ate a normal diet, it can be assumed that they do not differ from the general population. Since the additional administration of vitamin E led to a reduction in exercise-induced cellular damage, it can be assumed that athletes have an elevated vitamin E requirement.

Conclusions

Precise dose-finding studies for vitamin E supplementation have not been conducted to date. This does not appear to be very promising in view of the extreme difficulty experienced in determining marginal differences. The studies cited above were conducted with either more or less strongly pronounced vitamin E deficiency or with an additional vitamin E dosage of 200–1,200 IU vitamin E/day. Absorption studies have shown that vitamin E uptake from the intestines follows degressive absorption kinetics, i.e. the higher the administered dose, the lower the percentage utilization [28, 37]. From this point of view, it appears less sensible to substitute with single doses of more than 200 IU and daily doses of more than 600 IU. Exceptions would be fat malabsorption or transport dysfunctions, in which even higher doses are indicated.

There is no longer any doubt that the vitamin E requirement of athletes is increased in relation to their elevated turnover of energy and more intensive functional metabolism, along with the resulting problems of oxidative stress. For orientation purposes, a dose recommendation of 1 IU/kg body weight (normal persons) can be regarded as a suitable vitamin E intake, allowing for an increase under the special conditions of athletic activity. Thus, it makes sense to recommend a daily vitamin E intake of 100–200 IU for athletes, particularly those who do irregular training or start a new, more intensive phase of training. Since there is no toxicological risk within this dose range, but, in the case of low oxidative protection, there is a high risk of oxidative cellular damage, it would be bordering on malpractice not to point out the benefits of supplementation to the athlete. The above-mentioned investigations show that the uptake of vitamin E contained in the normal daily diet is clearly insufficient to minimize the negative effects of exercise-induced oxidative stress.

References

1   Berg A, Lehmann M, Keul J: Körperliche Aktivität bei Gesunden und Koronarkranken. Stuttgart, Thieme, 1986.
2   Demopoulos HB, Santomier JP, Seligman ML, et al: Free radical pathology: Rationale and toxicology of antioxidants and other supplements in sports medicine and exercise science; in Katch FI (ed): Sport, Health, and Nutrition. The 1984 Olympic Scientific Congress Proc, vol 2. Champaign, Human Kinetics, 1986, pp 139–190.
3   Elstner EF: Der Sauerstoff: Biochemie, Biologie, Medizin. Mannheim, BI Wissenschaftsverlag, 1990.
4   Jones DP: The role of oxygen concentration in oxidative stress: Hypoxic and hyperoxic models; in Sies H (ed): Oxidative Stress. London, Academic Press, 1985, pp 151–195.
5   Weineck J: Optimales Training. Erlangen, Perimed, 1988.
6   Aikawa KM, Quintanilha AT, De Lumen BO, et al: Exercise endurance-training alters vitamin E tissue levels and red-blood-cell hemolysis in rodents. Biosci Rep 1984;4:253–257.
7   Gohil K, Packer L, De Lumen B, Brooks GA, et al: Vitamin E deficiency and vitamin C supplements: Exercise and mitochondrial oxidation. J Appl Physiol 1986;60: 1986–1991.
8   Gohil K, Rothfuss L, Lang J, et al: Effect of exercise training on tissue vitamin E and ubiquinone content. J Appl Physiol 1987;63:1638–1641.
9   Lang J, Gohil K, Rothfuss L, et al: Exercise Training Effects on Mitochondrial Enzyme Activity, Ubiquinones and Vitamin E. Anticarcinogenesis and Radiation Protection. New York, Plenum Press, 1987, pp 253–527.
10 Packer L: Vitamin E, physical exercise and tissue damage in animals. Med Biol 1984;62:105–109.

11   Quintanilha AT: Effects of physical exercise and/or vitamin E tissue oxidative me-
      tabolism. Biochem Soc Transact 1984;12:403–404.
12   Carabello F, Liu F, Eames O, et al: Effect of viamin E deficiency on mitochondrial
      energy transfer. Fed Proc 1971;30:639.
13   Carabello FB: Role of tocopherol in the reduction of mitochondrial NAD. Can J
      Biochem 1974;52:679–688.
14   Fedelesova MP, Sulakhe PV, Yates JC, et al: Biochemical basis of heart function. IV.
      Energy metabolism and calcium transport in hearts of vitamin E deficient rats. Can
      J Physiol Pharmacol 1971;49:909–918.
15   Cormier M: Regulatory mechanisms of energy needs: Vitamins in energy utilization.
      Prog Food Nutr Sci 1977;2:347–356.
16   Schwarz K: Vitamin E, trace elements, and sulfhydryl groups in respiratory decline.
      Vit Horm 1962;203:42–52.
17   Schwarz K: The cellular mechanisms of vitamin E action: Direct and indirect effects
      of alpha-tocopherol on mitochondrial respiration. Ann NY Acad Sci 1972;203:42–
      52.
18   Grigoreva VA, Medovar FV: Studies on the components of the adenylic system in
      skeletal and cardiac muscles in experimental muscular dystrophy. Ukrain Biochem
      Zhur 1948;31:251–268.
19   Hummel JP: Oxidative phosphorylation process in nutritional muscular dystrophy.
      J Biol Chem 1948;172:421–429.
20   Lee YCP, King JT, Visscher MB: Role of certain minerals, vitamin E and other
      factors in the genesis of myocardial fibrosis in mice. Am J Physiol 1960;198:981–
      984.
21   Fitch CD, Sinton DW: A study of creatine metabolism in diseases causing muscle
      wasting. J Clin Invest 1964:43:444–452.
22   Sharman IM, Down MG, Sen RN: The effects of vitamin E and training on physi-
      ological function and athletic performance in adolescent swimmers. Br J Nutr 1971;
      26:265–276.
23   Sharman IM, Down MG, Norgan NG: The effects of vitamin E on physiological
      function and athletic performance of trained swimmers. J Sports Med 1976;16:
      215–224.
24   Shephard RJ, Campbell R, Pimm P, et al: Vitamin E, exercise, and the recovery
      from physical activity. Eur J Appl Physiol 1974;33:119–126.
25   Talbot D, Jamieson J: An examination of the effect of vitamin E on the performance
      of highly trained swimmers. Can J Appl Sports Sci 1977;2:67–69.
26   Cureton TK: The physiological effects of wheat germ oil on humans in exercise.
      Springfield, Thomas, 1972.
27   Clausen D: The combined effect of aerobic exercise and vitamin E upon cardiores-
      piratory endurance and measured blood variables; masters thesis, University of
      Wyoming, 1971.
28   Nagawa T, Kita H, Aoki J, et al: The effect of vitamin E on endurance. Asian Med J
      1968;11:619–633.
29   Kobayashi Y: Effect of vitamin E on aerobic work performance in man during acute
      exposure to hypoxic hypoxia; thesis, University of Albuquerque, 1974.
30   Simon-Schnass I, Pabst H: Influence of vitamin E on physical performance. Int J
      Vitam Nutr Res 1988;58:49–54.

31   Böhlau V, Böhlau E: Clinical investigations on the influence of vitamine E on the efficiency of the body; in Shimazono N (ed): International Symposiun on Vitamin E: Hakone, Japan, 8/9 September 1970. Tokyo, Kyoritsu Shuppan, 1970, pp 238–249.

32   Packer L, Reznick AZ, Simon-Schnass I, et al: The significance of vitamin E for the athlete; in Vitamin E: Biochemistry and health implications. New York, Marcel Dekker, 1993, pp 465–471.

33   Dillard CJ, Dumelin EE, Tappel AL: Effect of dietary vitamin E on expiration of pentane and ethane by the rat. Lipids 1977;12:109–114.

34   Kanter HM, Nolte JA, Holloszy JO: Effects of an antioxidant supplement on expired pentane production following low and high intensity exercise. Med Sci Sports Exerc 1991;22:86.

35   Apple FS, Rhodes M: Enzymatic estimation of skeletal muscle damage by analysis of changes in serum creatine kinase. J Appl Physiol 1988;65:2598–2600.

36   Simon-Schnass I, Korniszewski L: The influence of vitamin E on rheological parameters in high altitude mountaineers. Int J Vitam Nutr Res 1990;60:26–34.

37   Böhles E: Resorptionsdynamik unterschiedlicher Vitamin-E-Mengen bei gesunden jungen Männern. VitaMinSpur 1988;3:134–136.

38   Simon-Schnass I, Reimann J, Böhlau V: Vitamin-E-Therapie: Zur Frage der Resorption von Vitamin E. Notabene Medici 1984;14:793–794.

39   Weinstock IM, Schoichet I, Goldrich AD, et al: The effect of vitamin E deficiency on the oxidation of tricarboxylic acid cycle intermediates. Arch Biochem Biophys 1955; 57:496–505.

Dr. Irene Simon-Schnass, Hermes Arzneimittel GmbH, Georg-Kalb-Strasse 5–8, D-W–8023 Grosshesselohe (FRG)

Simopoulos AP, Pavlou KN (eds): Nutrition and Fitness for Athletes.
World Rev Nutr Diet. Basel, Karger, 1993, vol 71, pp 154–162

Principles of Athletes' Nutrition in The Russian Federation

Victor A. Rogozkin

Research Institute of Physical Culture, St. Petersburg, Russia

Introduction

Nutrition plays a leading role among numerous environmental factors and conditions which constantly influence human beings. In the process of nutrition, food changes from being an external factor to an internal one, and its components in a succession of enzymatic reactions are transformed into the energy of physiological functions and structural elements of the human body. The influence of food is predominant in ensuring optimal growth and development of the human organism, as well as in its work capacity and in adaptation of the effects of the various environmental factors. In the end nutrition exerts a principal influence upon humans in being active and having a prolonged life. The up-to-date concepts related to nutrition include a wide complex of biochemical and physiological mechanisms of food assimilation, and among them the processes of digestion and transformation of nutritional substances in the organism's inner workings. On the basis of the analysis of these processes the ideas of the needs of the human organism and indispensable nutritional substances are established [1].

In this article we are going to briefly consider the organization of human rational nutrition as an indispensable condition in ensuring a standard level of physical work capacity in the process of everyday activities. Then we will discuss the fundamental principles of athletes' nutrition. Finally, the possibilities of using special nutritional supplements at various stages of athletes' preparation and their participation in competitions will be shown.

Table 1. Nutrients required by man

Amino acids	Elements	Vitamins
Established as essential		
Isoleucine	Calcium	Ascorbic acid
Leucine	Chlorine	Choline
Lysine	Copper	Folic acid
Methionine	Iodine	Niacin
Phenylalanine	Iron	Pyridoxine
Threonine	Magnesium	Riboflavin
Tryptophan	Manganese	Thiamine
Valine	Phosphorus	Vitamin B_{12}
	Potassium	Vitamins A, D, E, and K
	Sodium	
	Zinc	
Probably essential		
Arginine	Fluorine	Biotin
Histidine	Molybdenum	Pantothenic acid
	Selenium	Polysaturated fatty acid

Principles of Human Rational Nutrition

The balance of basic nutritional substances and the right diet are the basis of human rational nutrition which ensures an appropriate level of vital functions and physical work capacity. First of all it is necessary to answer the following question: 'What should be taken as principle in determining the amount, composition and quality of food products?' To answer this question it is necessary to compare human needs in terms of nutritional substances with the chemical composition of food products. In our country the up-to-date ideas about the food needs of human beings have been developed following the investigations of Pokrovsky [2] and Ugolev [3] who established the concepts of rational nutrition.

Table 1 gives some basic components of food which are indispensable for a grown human [1]. These are 10 amino acids, 13 vitamins and 14 mineral elements. The disturbance in satisfying the needs of an organism concerning these indispensable nutritional substances leads as a rule to diseases of nutritional insufficiencies which are manifested in disturbing the processes of metabolism and lead to a decrease of human physical work capacity.

The basic components on which all the subsequent calculations are based is the value of food ration energy values, which is in close connection with determining human physical power inputs in real conditions of vital functions. The power inputs of humans include four components: basic exchange, physical activity, growth and lactation, and special dynamic effect of food. In order to determine the human power inputs different methods of direct and indirect calorimetry are used [4]. During the last few years the method of heart rate integration for evaluating the human power inputs by performing physical exercises has been used. In this case a compact recorder of heart rate – a sports tester and a personal computer – are used. Our wide experience gained at the Institute shows that the application of sports tester for recording the heart rate by performing physical exercises enables us to reliably evaluate the human power inputs during training.

On the grounds of these findings it is possible to calculate the power value of training and form some notions about human daily energy intake [5].

Depending on the value of real power inputs the calorie content of daily food and the indispensable amount of proteins, fats, carbohydrates, vitamins and minerals are calculated. The character and nature of nutritional substances used in the composition of all food products is of great importance for organizing the human rational nutrition.

Fundamental Principles of Athletes' Nutrition

The needs of an athlete in energy and nutritional substances essentially differ depending first of all on the kind of sport and the amount of work performed [6]. If one compares the food allowance of a girl gymnast or a flyweight boxer with that of a track-and-field athlete or a weight lifter having super heavy weight, sharp differences in main nutritional components will manifest themselves at once. However, even without having recourse to such contrasting comparisons it should be taken into consideration the fact that the nature of athletes' nutrition is directly associated with metabolic processes occurring in an organism by practicing one or another kind of sports. Therefore, the main differences among the athletes according to their needs in energy and nutritional substances are bound with the specific nature of sport (table 2).

The need of athletes for proteins is in many ways defined by the specific nature of one or another sport, trend of training process, amount and

Table 2. Dietary recommendations for athletes

Sport	Energy, kcal		Carbo-hydrate	Protein % cal	Fat
	male	female			
Track sprint, hurdles, throwing, jumping, weight lifting, swimming	3,500–4,500	3,000–4,000	52–53	17–18	30
Distance running, rowing, cycling, skiing, cross-country skating	5,500–6,000	5,000–5,500	60–61	14–15	25
Archery, gymnastics, figure skating, fencing, ski jumping	3,500–4,500	3,000–4,000	57	15	28
Wrestling, judo, boxing	4,500–5,500	4,000–4,500	53–54	17–18	29
Basketball, field hockey, football, handball, hockey, tennis, volleyball	4,500–5,000	4,000–4,500	55–58	15–17	28

intensity of physical exercise. As a rule we recommend to include 1.4–2.0 g of protein per 1 kg of body weight. Depending on the kind of sport the normal fat intake may be 1.7–2.4 g/kg of body weight. The need of athletes for carbohydrates is in close connection with power inputs during the training and competitions. The norm of carbohydrate uptake makes up 8–14 g/kg of body weight. The content of basic nutritional substances in athletes' food allowance makes up the following: proteins, 15–16%; fats, 25–26%; and carbohydrates, 58–60% of total calories. The need of athletes for vitamin supplements, minerals, and nutritional fibers requires a special discussion.

For about 20 years we have been studying the problem of athletes' nutrition, and during this period we carried out a number of investigations and held several international conferences and symposia. We have collected information that is included in the book *Athletes' Nutrition* [7] which was published in our country in 1989, and very soon it became a best-seller.

In this book we are expounding the following fundamental principles of athletes' nutrition.

(1) Provide the human organism with the necessary amount of energy that is proportional to an intake that is necessary to the process of performing physical exercise.

(2) Keep nutritional balance as applied to certain kinds of sport and the intensity of physical exercises including the distribution of nutritional substances in terms of energy value that will alter depending on the periods of preparation for competitions. It applies equally to the balancing of amino acids forming a part of protein products and by the amount of basic nutritional substances, nutritional fibers, vitamin supplements and micronutrients, as well as keeping of optimal proportions in fatty acid composition.

(3) Selection of adequate nutritional forms (products, nutritional substances and their combinations) in the period of intensive and prolonged physical exercise, immediate preparation for competitions, the competitions themselves and the subsequent recovery.

(4) Use of nutritional substances for an activation and regulation of intracellular metabolic processes in diverse organs and tissues.

(5) Creation of a necessary metabolic background with the help of nutritional substances for the biosynthesis and realization of hormonal actions regulating the inclusion of metabolic reactions (catecholamines, prostaglandins, corticosteroids, cyclic nucleotides and others).

(6) Maintain a variety of foods by using a variety of products and by applying various methods of their culinary treatment in order to optimally ensure the organism with all necessary nutritional substances.

(7) Include in food allowances some biologically valuable foodstuffs with dishes that can be quickly digested and which do not burden the alimentary canal.

(8) Use of nutritional factors for raising the velocity of the development of muscular mass and the increasing strength, as well as for regulating the body weight according to athletes' weight class.

(9) Individualization of athletes' anthropometric, physiologic and metabolic characteristics, the condition of his digestive system, his personal tastes and habits.

For many years these principles of forming a balanced nutrition have been widely used for the preparation of athletes in various climatic zones and geographic locations of our country.

The use of these theses by creating the food allowances of athletes enables to better satisfy the human organism's metabolic requirements and to maintain a high level of physical work capacity.

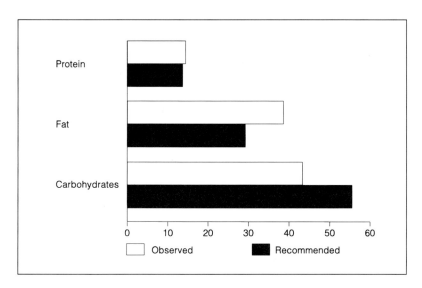

Fig. 1. Distribution of protein, fat and carbohydrates (in energy %): comparison between observed and recommended daily allowances for elite athletes.

It should not be forgotten that in the conditions of unbalanced nutrition, due to lack in indispensable food factors or to disturbances in the ratio of fundamental nutritional substances, the normal realization of metabolic processes in an organism may be altered, and the latter leads to reduction in physical work capacity.

The following are some typical examples of the violation of rational nutrition basic formula; these examples are often seen in those practicing sports.

Evaluation of real athletes' nutrition revealed an essentially larger consumption of fats irrespective of athletes' age, kind of sports and the stage of preparation (fig. 1). At the same time an insufficient consumption of carbohydrates whose content is lower than it is necessary in daily ration was observed [7]. A scanty consumption of vitamins in food is shown rather often in athletes, especially in the winter-spring period. As a rule, the deficiency of vitamins C, B_1, B_2 and PP consumption is observed.

The third example is associated with an insufficient consumption of mineral substances, and first of all iron. The development of iron-deficiency anemia in athletes is usually associated with the disturbance of balance between the assimilation of iron that is obtained from food and the need of the organism. Among the athletes practicing cycling, racing, skiing, skating, long-distance running and walking various forms of iron-deficiency anemia are widespread.

The maintenance of a high level of human physical work capacity under the conditions of ordinary life is possible by providing rational nutrition and ensuring that essential nutrients are not lacking [8]. At the same time, it is necessary to obtain some more exact data concerning athletes' metabolic needs related to fundamental nutritional substances and energy depending on professional activities typical for different sports including training and competition.

Special Nutritional Supplements for Athletes

Now let us consider some possibilities of using nutritional factors to increase physical work capacity of athletes in the period of preparation and during competitions. Numerous investigations have provided data on the biochemical mechanisms involved in muscular activity in terms of energy intake and specific nutrients [6, 9–14]. The results of these investigations indicate that there exists a real opportunity to improve athletes' physical work capacity which is associated with an increase of carbohydrate energy sources (first of all glycogen) in skeletal muscles and liver by ingesting food products with a high content of easily assimilated carbohydrates.

It is impossible to achieve daily power inputs in athletes (up to 6,000–7,000 kcal) by means of the usual foodstuffs, even those possessing a high biological value. The existing needs in vitamin supplements in athletes are not always compensated by traditional nutrition, owing to the fact that the intensity and duration of everyday training do not leave the time for a standard assimilation of basic food in the gastrointestinal tract to supply all the organs and tissues with essential nutrients. Such alterations in metabolism lead to a reduction in velocity of energy and a slowing of tissue recovery of the organism; it has an effect on sports work capacity and impedes the increase of attainments in sports. Such merits of special nutritional supplements as their high nutritional density, a pronounced food trend, the homogeneity, a variety of convenient forms of transport and cooking, good flavoring and reliable hygienic qualities enable them to be successfully applied in the nutritional plan for athletes.

In sports practice special nutritional supplements are used for: (a) nutrition for distance swimming and between trainings; (b) speeding up of the processes of recovery after the trainings and competitions; (c) regulation of water-salt exchange and heat regulation; (d) correction of body weight; (e) directed development of muscular mass (body building); (f) reducing the volume of daily rations during competitions; (g) change of daily ration qualitative orientation depending on the trend of training loading or by the preparation for competitions; (h) individualization of nutrition, especially under the conditions of high nervous-emotional stresses; (i) urgent correction of unbalanced daily rations, increase of the number of meals under the conditions of multiple trainings and competitions.

All the nutritional supplements may be divided into three classes: protein and compound mixtures, carbohydrate-mineral beverages and vitamin-mineral complexes. The great part of them represent dry granular mixtures, tablets and more seldom confectionery.

Today about 10–15 special supplements of various nutritional factors intended for athletes are produced in our country.

Conclusion

The correspondence of nutrition nature to metabolic alterations caused by muscular activity in many ways defines the development of the processes of athletes' adaptation to performance training and competitive physical exercise. The various nutrients actively influence the metabolic processes in an organism and increase the sports work capacity, and accelerate the processes of its recovery in the period of rest after training and competitions. It should not be forgotten that a contrary effect is also possible, that is when under the conditions of unbalanced nutrition and by a lack of indispensable nutritional components the standard proceeding of metabolic processes can be disturbed. It will lead to a reduction of athletes' work capacity.

So, what is it necessary to do in order to avoid general mistakes in organizing a rational nutrition for those involved in physical training programs?

It is necessary to have an elemental knowledge concerning the basic foodstuffs, their calorie content and the composition of nutritional substances, as well as the organization of nutrition and diet. All this insistently calls for a broad instruction of the population on the principles and rules of rational nutrition.

References

1   White A, Handler P, Smith E, et al: Principles of Biochemistry. Moscow, Mir, 1981, vol 3, pp 1703–1717.

2   Pokrovsky AA: The role of biochemistry in development science of nutrition. Science (Moscow) 1974:7–15.

3   Ugolev AM: Theory for equivalent nutrition. Klinicheskaja Med 1986;14:15–24.

4   Lamb DR: Physiology of Exercise. New York, Macmillan, 1984, pp 99–113.

5   Rogozkin VA, Zagrantsev VV, Pshendin AI: Measurement of energy expenditure during exercise (abstract). XXXI Int Congr Physiol Sci, Helsinki, 1989, p 550.

6   Rogozkin VA: Some aspects of athlete's nutrition; in Parizkova J, Rogozkin V (eds): Nutrition, Physical Fitness and Health. International Series on Sport Science. Baltimore, University Park Press, 1978, pp 119–123.

7   Rogozkin VA, Pshendin AI, Shishina NN: Athletes' Nutrition. Moscow, FIS, 1989.

8   Rogozkin VA: Nutrition and physical fitness: present and future; in Feldkoren B (ed): Nutrition and Physical Fitness. Leningrad, 1991, pp 3–11.

9   Rogozkin VA: The role of low molecular weight compounds in the regulation of skeletal muscle genome activity during exercise. Med Sci Sport 1976;8:74–79.

10 Rogozkin VA, Pshedin AI, Shishina NN: Metabolic adaptation to different diets and physical exercise. Am J Clin Nutr 1989;49:1127.

11 Williams C: Diet and endurance fitness. Am J Clin Nutr 1989;49:1127.

12 Coyle EF: Carbohydrate feedings; effect on metabolism, performance and recovery; in Brouns F (ed): Advances in Nutrition and Top Sport. Med Sport Sci. Basel, Karger, vol 32, 1991, pp 1–14.

13 Hargeaves M: Carbohydrates and exercise. J Sport Sci 1991;9:17–28.

14 Scherman WM: Carbohydrate feedings before and after exercise; in Lamb DR, Williams MH (eds): Perspectives in Exercise Science and Sports Medicine. Ergogenics: Enhancement of Performance in Exercise and Sport. Indianapolis, Benchmark, vol 4, 1991, pp 1–27.

Prof. V.A. Rogozkin, Research Institute of Physical Culture, Avenue Dynamo 2, 197402 Saint Petersburg (Russia)

Simopoulos AP, Pavlou KN (eds): Nutrition and Fitness for Athletes.
World Rev Nutr Diet. Basel, Karger, 1993, vol 71, pp 163–194

Poster Abstracts

Obesity Indexes I and II

G.P. Aggelikakis
TEI/Nutrition Department, Sindos, Thessaloniki, Greece

We suggest two Obesity Indexes I and II as simple criteria for the nutritional status including normal state, obesity and malnutrition. Both of them are functions of three variables: height, weight and wrist circumference. Indexes I and II were calculated by linear regression analysis, the first one from height-weight tables and the second one from the relationship $W = (BMI): H^2$ which for a given value of BMI is linear ($r = 0.9993$) in the range of $H \geq 1.5$ m. After wrist circumference correction, the extracted equations were transformed to give a theoretical value equal to zero for normal persons. Index I consists of two equations (one for each sex) and Index II is one equation for two sexes. The proposed indexes were tested by application on two small random samples (50 men 28.1 $\pm$ 8.9 years and 65 women 25.7 $\pm$ 6.8 years). The application showed that Obesity Index II is a criterion more sensitive and suitable for the nutritional status than Obesity Index I. The observed range of Obesity Index II was as follows:

	Men	Women
Malnutrition	< -1.7	< -3.9
Normal state	-1.1 to 4.9	-4.3 to 3.7
Slight obesity	4.4 to 9.6	3.6 to 10.7
Obesity	> 11.7	> 11

Does HDL Injection Reduce the Development of Serum Hyperlipidemia and Progression of Fatty Streaks in Cholesterol Fed Rabbits?

J. Beitz[a], A. Beitz[a], A.Y. Misharin[b], K.-E. Pawleski[a], H.-J. Mest[a]
[a] Martin Luther University, Halle, FRG; [b] Academy of Medical Sciences, Moscow, Russia

Serum hyperlipidemia is one of the major risk factors for development of atherosclerosis, probably as a consequence of unbalanced cholesterol influx-efflux in the tissues. The antiatherogenic action of high-density lipoprotein (HDL) includes that HDL serves as a scavenger for exchangeable tissue cholesterol. The aim of the present study was to investigate whether injections of exogenous HDL i.v. were able to reduce the development of serum hyperlipidemia or progression of fatty streaks in cholesterol-fed rabbits. Therefore, we fed 3 groups of New Zealand rabbits with a 0.5% cholesterol-rich diet for 8 weeks and injected the following solutions intravenously: group 1 (control): weekly 2 times 3 ml saline; group 2: weekly 2 times 5 mg human HDL (quantified as protein), dissolved in 3 ml saline; group 3: weekly 2 times 5 mg rabbit HDL (quantified as protein), dissolved in 3 ml saline. The animals of group 2 had lower serum cholesterol levels than the rabbits of group 1 ($p < 0.05$), but the surface of intima covered with fatty streaks was the same as in group 1. On the contrary, the serum cholesterol level in rabbits of group 3 was the same as in group 1 during the whole experimental period, but the surface of aorta covered with fatty streaks was significantly lower ($p < 0.05$).

The results of this study support the hypothesis of an antiatherogenic action of HDL, but this action seems to be independent of the influence of HDL on the total serum cholesterol concentration.

The Nutritional Value of Daily Food Rations of Polish Soldiers Serving in United Nations Disengagement Observer Force (UNDOF) in Syria

J. Bertrandt, A. Klos, M. Miklasz, J. Szczesniak
Military Institute of Hygiene and Epidemiology, Warsaw, Poland

The quality of nutrition of Polish soldiers serving in UNDOF in Syria was assessed. On the basis of theoretical analysis of 50 daily food rations, the amount of energy, protein, fat, carbohydrates, vitamins and mineral elements were calculated. Results were compared with norms for nutrition of Polish soldiers. It was found that nutritional values of daily intake of the recommended norms for energy, protein, Ca, Fe and riboflavin were accomplished. Amounts of fat and vitamin C were higher and carbohydrates, especially cellulose, P and Mg were lower than the Polish norms.

Effects of Salmon and Wheat Germ Oil Administration on Fatty Acid Composition of Erythrocyte Membranes in American Football Players

P.L. Biagi[a], *C. Spano*[a], *L. Venturoli*[b], *E. Turchetto*[a]
[a]Universita di Bologna, and [b]American Football Team, Bologna, Italy

Introduction. American football needs a high level of mental and physical exertion, aerobic and anaerobic. Salmon and wheat germ oil give fatty acids of the n–3 and n–6 series. We studied the influence of oil administration on athletic performance and membrane composition.

Materials and Methods. We obtained blood samples of 30 players, divided in two groups, before and after salmon oil (3 g/day) and wheat germ oil (Golden Products) (6 g/day) administration. The administration continued for 4 weeks. We studied fatty acid composition of erythrocyte membranes because it is a model of the long-term nutritional influence. Erythrocyte membranes were extracted with the Dodge method, their lipids with the modified Folch method. Methyl esters were esterified according to the Stoffel method. High-performance gas chromatographic analysis was performed using a Carlo Erba instrument HRGC model 5160 equipped with fused silica column mod. SP 2340 (Supelco) 30 m $\times$ 0.25 mm ID. Column temperature was programmed from 160 to 210 °C, with a gradient of 8 °C/min, gas carrier flow rate was 2 ml/min. Statistical analysis was performed using paired t test.

Results. In the group treated with wheat germ oil there was a decrease of 16:1n–7 ($p < 0.05$), 20:5n–3 and 22:5n–3 ($p < 0.001$); an increase of 18:3n–6 ($p < 0.05$); and a trend to increase of 18:2n–6. In the salmon oil group there was a decrease of 17:0 ($p < 0.05$) and 22:5n–3 ($p < 0.001$); a trend to increase of 22:6n–3.

Discussion. Results suggest that the oil administration influenced fatty acid composition of erythrocyte membranes, but this effect was not more significant because athletes followed a trial of aerobic practices; in fact, fatty acids administered were utilized as an energy source during exertion.

Conclusion. We can say that oil administration influences the athletic performance and secondary, the acidic pattern. We are waiting for better results after another supplementation and anaerobic trial of practices.

A special acknowledgement of Golden Products, Italia for oil supplies, and to A.S. Phoenix American Football Team for the collaboration. Thi work was supported by ISEF (60%), E.T. and P.L.B.

Effect of Body Composition on Work Capacity of the Human Evaluated by Mechanical Efficiency

V. Bunc, H. Heller
Charles University, Prague, Czechoslovakia

The utilizable sources of energy of the human organism are relatively restricted whereby interindividual differences are not substantial, while differences in physical performance are extreme. The capacity to transform chemical energy into mechanical work, sometimes described also as the work capacity of the organism, may be evaluated by means of mechanical efficiency (ME). We assessed ME on a treadmill in men with varying amounts of training (including untrained ones) who differed significantly as to the percentage of body fat (3.8–17.2%) which we assessed by means of ten skinfolds. The highest ME values, and thus the best predispositions for physical activity (that are types of walking and/or running) were found in trained sportsmen (endurance runners – 32.4 ± 4.1%); the lowest ME were found in untrained middle-aged men (16.9 ± 4.6%). The authors found a close negative correlation of ME (ME = $-1.133 \cdot$ % fat + 35.634, r = -0.336, p < 0.001) and maximal oxygen uptake per body mass (VO_2 max$\cdot$kg^{-1} = $-2.741 \cdot$ % fat + 85.185, r = -0.349, p < 0.001) and values of body fat. With an increasing percentage of body fat the values of ME decline, the prerequisites for effective transformation of chemical energy of the organism into mechanical work are reduced, the work capacity of the human declines and decreases the possibility of effective muscle work.

Simple Assessment of Aerobic Fitness with a Two-Kilometer Running Test

V. Bunc
Charles University, Prague, Czechoslovakia

Physical activity is becoming an integral part of everyday life. In order to modify and control this activity we need simple and readily available criteria for the control of effect of training, i.e. physical fitness. One of the possible approaches to resolve this problem is to use motor tests. The most frequently used method for aerobic fitness proposed by Cooper, based on measurement in the American population, may involve, when used under European conditions, inaccuracy of estimates of aerobic fitness and be associated with some difficulties regarding the 12-min period of the test. The author submits, therefore, tables for the estimation of aerobic fitness and physical performance elaborated with regard to Czech and European population standards and general relationships between velocity of running and energy required for this activity expressed indirectly by oxygen consumption. The basic element of evaluation in the field is the mean velocity of movement on a 2-km distance. The tables were prepared for men and women aged 14–65 years and make it possible to estimate poor, good and excellent levels of physical performance. The reliability of assessment of maximal oxygen uptake, and thus also the estimate of aerobic fitness, and therefore the physical fitness level varies around 15%.

Relevance of Designer Foods in Health and Fitness

Ritva R. Butrum, Herbert F. Pierson
Food Phytochemical Research Institute, Rockville, Md., USA

Food phytochemicals (FPs) are defined as low molecular weight organic compounds in edible plants. FPs confer survival properties to plants and induce subtle metabolic influences on detoxification pathways in mammals. The potential that many of these compounds, abundant in the food supply, are important to the integrity of physical well-being and resistance to chronic diseases is recognized. The development of Designer Foods focuses on health-promoting potential of fruits, vegetables and grains, i.e. edible plants. These Designer Foods are contrived and formulated to be standardized in composition with naturally occurring levels of known health-promoting constituents. Presently more than 20 NCI (National Cancer Institute) funded research projects are nearing completion. Their scope includes composition of various plant constituents, evaluation of their tolerance and absorption, and measurement of their metabolic modulatory influences, i.e. modulation of prostaglandins, xenobiotics and steroid hormones and other key pathways.

The Effects of Diet and Exercise on Bone Mineral Density

Ann M. Fehily, W. Evans, M.L. Burr
MRC Epidemiology Unit and University Hospital of Wales, Cardiff, UK

A study of forearm bone mineral density was conducted in 371 subjects aged 20–23 years. The subjects had been identified 14 years earlier as being at high risk of inadequate nutrition and had then participated in a randomized trial of supplementary milk. Bone mineral density was associated positively with body weight ($p < 0.01$) in all subjects, and with current calcium intake ($p < 0.05$), vitamin D intake ($p < 0.01$), and sport during adolescence ($p < 0.01$) in women. It was associated negatively with alcohol intake ($p < 0.05$) in men. Nonsignificant associations were found with milk supplementation (positive) in all subjects, with manual occupation (positive) in men, and with parity (negative) in women. Diet and exercise are determinants of bone mineral density in early adult life.

Regulation of Phosphorylase Activity and Glycogenolysis in Muscle during Short-term Exercise in Humans

Dimitrios Chasiotis
Hellenic Sports Research Institute, Athens, Greece

Phosphorylase activity and the rate of glycogenolysis in muscle have been investigated in humans using the needle biopsy technique. The fraction of phosphorylase *a* in resting muscle was 23% but the rate of glycogenolysis was zero. Epinephrine infusion increased cAMP and phosphorylase *a* in muscle by 3- and 4-fold, respectively. The maximum rate of glycogenolysis was 11 mmol kg^{-1}·dm·min^{-1} which is 5–10% of the Vmax *a* determined in vitro.

Isometric contraction to fatigue at 66% MVC increased phosphorylase *a* 2.5 times and the rate of glycogenolysis was about 50 mmol·kg^{-1}·dm·min^{-1}. No significant difference in the rate of glycogenolysis in muscle was observed after isometric contraction to fatigue without and with epinephrine infusion. Continuous and intermittent isometric contractions induced by electrical stimulation increased phosphorylase a 3.5 times and the rate of glycogenolysis varied between 70 and 90 mmol $kg^{-1} \cdot dm \cdot min^{-1}$.

The results in the present studies can be explained on the basis of substrate regulation of phosphorylase activity provided that the P_i is present in a limiting amount at the active site as phosphorylase in muscle at rest. It is concluded that increased rate of glycogenolysis in muscle is a function both of phosphorylase in the a form and the availability of P_i at the active site of the enzyme.

Physiological Effects of Fluid, Electrolytes, and Substrate Replacement on Athletic Performance in Hot Environments

J.D. Chen, Q.H. Yang, Z.Y. Yang, W.M. Weng and co-workers
Beijing Medical University, Beijing, China

In order to delay the onset of fatigue induced by heat stress and improve athletic performance, we designed a hypo-osmotic and isotonic fluid and examined its effects. Subjects were eight athletes of Beijing Physical Education Institute; age: 19–23 years; weight: 63.3 ± 4.4 kg; height: 1.76 ± 0.03 m. Exercise was conducted on a Monark Ergometer in a climate chamber (T: 32.6 ± 0.3 °C, RH: 70.4 ± 4.2%), physical load was adjusted for the first 7–10 min to keep the HR steady at 150 beats/min and remained constant thereafter. Subjects stopped exercising when their HR was > 162 beats/min. In every experiment, 2 subjects exercised at the same time, one was given the experimental fluid and the other a same colored and flavored placebo within fixed times. The experiment was repeated, but the same amount of fluid received by the subjects was reversed. Venous blood samples were obtained before, after 1-hour cycling and immediately after exercise for analysis of Hb, Hct, glucose, corticol, lactate and electrolytes. Sweat volume, skin and core temperature, HR, body heat accumulation rate, endurance time and work output were monitored.

Main Results. (1) Sweat rate was significantly lower in athletes with the experimental fluid replacement than that of the placebo: 15.0 ± 3.4 vs. 16.9 ± 4.1 ml/min (p < 0.05). (2) Blood glucose levels of athletes under the experimental fluid replacement decreased within a narrow range but there was no hypoglycemia after exercise, while half of the subjects suffered from hypoglycemia under placebo replacement. (3) There was significant difference between the increase of core temperature of subjects under experimental and placebo fluid replacement, 1.27 ± 0.49 vs. 1.44 ± 0.54 °C (p < 0.01). The heat accumulation rate was substantially lower in subjects under experimental fluid supplements, 1.335 ± 0.744 vs. 1.745 ± 0.808 kJ/m²/min (p < 0.05). (4) Endurance time of athletes under experimental fluid replacement was significantly longer and their work output was obviously greater as well. Endurance time was 107.0 ± 0.19 vs. 92.5 ± 14.9 min (p < 0.01). Work output was 684.0 ± 113 vs. 597 ± 80 kJ (p < 0.05). These results show that the physiological responses were improved and especially the heat stress alleviated in athletes under fluid, electrolytes and substrate replacement which led to better athletic performance.

Effect of Physical Activity Level on the Functional Capacity of Six Muscle Groups in Women Aged 25–74 Years

C.B. Christ[a], *R.A. Boileau*[a], *M.H. Slaughter*[a], *R.J. Stillman*[a], *J.R. Johnson*[b]
[a]Universiy of Illinois, Urbana, Ill.; [b]Longwood College, Farmville, Va., USA

The influence of habitual physical activity on the functional capacity of skeletal muscle in women is uncertain; hence, this study was designed to investigate the effect of physical activity level (PAL) on two parameters of muscle function (MF): maximal force (MFR) and maximal rate of force increase (MR) in six muscle groups (MG) representing contrasts of size, proximity, extremity and usage. Physical activity level was assessed by questionnaire, and the mean of three independent ratings was used as the criterion rating to classify each subject as low (LO, n = 38), moderate (MOD, n = 50), or high (HI, n = 58) activity. The inter-rater reliability for PAL assessment ranged from r = 0.54 to r = 0.68 (mean = 0.59). Repeated measures (2 days $\times$ 3 trials of MFR and MR were obtained for the finger flexors (FF), thumb extensors (TE), forearm flexors (FAF), forearm extensors (FAE), dorsi flexors (DF) and plantar flexors (PF) utilizing a linear voltage differential transducer interfaced with an IBM microcomputer. When the influence of age and body weight on MF was statistically controlled, significant ($p < 0.05$) differences in MFR and MR were observed among PALs for all of the muscle groups tested except of the PF (MFR, MR) and the FAF (MR). Analysis of covariance yielded the following adjusted means:

MG	Maximal force, kg			Maximal rate, kg/s		
	LO	MOD	HI	LO	MOD	HI
FF	26.8	27.7	30.2	176.3	182.7	207.7
TE	3.7	4.2	4.4	18.1	21.5	24.5
FAF	15.6	17.8	18.5	86.6	100.1	101.5
FAE	6.5	7.3	8.3	33.3	38.0	49.9
DF	21.8	24.4	25.3	120.7	133.0	141.1
PF	75.9	83.7	84.4	227.5	252.2	269.0

Both muscle function variables were typically greater in the HI activity group when compared to the MOD and LO groups. Although significant differences were not observed between the MOD and LO activity groups, the muscle function of the MOD group was slightly greater than the LO group. These results suggest that a physically active lifestyle may be advantageous in maintaining the functional capacity of skeletal muscle in women.

Effects of Long-Term Physical Training in Adult Men

Hiroshi Ebashi[a], Hidetaro Shibayama[b]
[a] Meiji Life Foundation of Health and Welfare, Hachioji;
[b] National Institute of Fitness and Sports, Kanoya, Japan

Effects of chronic regular physical training upon the respiratory-circulatory function and blood constituents were studied. Two healthy men, 32 and 38 years of age at the beginning of training, served as the subjects. The exercise for training was 20 min running on a treadmill at 3° inclination. The exercise intensity was quivalent to 66% of maximal oxygen consumption (VO_2 max) in each subject and was readjusted periodically. The exercise was performed once a day, 5 days a week and was continued for approximately 17 years.

Hemoglobin (Hb) and hematocrit (Hct) were decreased gradually after the third year. In one subject Hb and Hct dropped to 9 g/dl and 27%, respectively, at the ninth year. Both Hb and Hct levels were improved following oral iron supplementation, but still lower than the pretraining levels. Serum triglyceride concentration decreased significantly approximately 90–120% after 1 year and relatively constant levels have been maintained for 17 years. Blood lactate concentration after exhaustive exercise decreased in response to training but no correlation was seen with training duration. Blood glucose level after exhaustive exercise was unchanged during training.

Heart rate at rest decreased by 5–6 beats/min approximately 3 months after initiation of training. Resting heart rates have been maintained at lower levels than pretraining. Heart rate and respiratory frequency during exercise tended to be reduced following training. However, no correlation between these parameters and training period was observed. The VO_2 max decreased by 20–30% after 17 years relative to the pretraining level.

It is suggested that chronic jogging training aimed at maintenance and improvement of health and physical fitness levels may be useful to increase the efficiency of oxygen supply during exercise, but not necessarily the physical activity.

Thiamin, Riboflavin, Pyridoxal, Magnesium, Iron and Zinc Status in Females during a 24-Week Fitness-Type Exercise Program

Mikael Fogelholm[a], Juha Laakso[b], Inkeri Ruokonen[b]
[a] University of Helsinki, and [b] MILA Laboratory, Helsinki, Finland

The objective of this study was to investigate whether a 24-week progressive fitness-type exercise program affected vitamin and mineral status in females. In addition to blood indices of thiamin, riboflavin, vitamin B_6, magnesium, iron and zinc status, data on physical activity and nutrient intake were collected by repeated 4-day recording. Subjects in exercise (n = 21) and control (n = 18) groups were female university students, aged 18–33 years.

In the beginning, estimated total daily energy expenditure in exercise (8.1 ± 0.1 MJ) and control subjects (8.1 ± 0.2 MJ) was similar. In the exercise group, energy expenditure

Table 1. Daily micronutrient intake: mean results from 3 × 4-day food records; estimated percentage of inadequate dietary intakes, obtained from probability analyses, are shown in parentheses

	Exercise		Control	
Thiamin, mg	1.3 ± 0.1	(4)	1.3 ± 0.1	(3)
Riboflavin, mg	1.9 ± 0.1	(2)	1.8 ± 0.1	(3)
Vitamin C, mg	157 ± 14	(0)	133 ± 10	(0)
Calcium, mg	1,060 ± 77	(8)	980 ± 50	(4)
Magnesium, mg	318 ± 16	(14)	311 ± 14	(7)
Iron, mg	13.7 ± 0.7	(22)	13.4 ± 0.6	(28)
Zinc, mg	11.2 ± 0.6	(21)	10.7 ± 0.5	(27)

increased over the study period to 8.9 ± 0.2 MJ/day. Energy intake in exercise subjects did not vary significantly, nor did it differ from controls. Vitamin and mineral intake was not different between the two groups (table 1).

Erythrocyte (E) transketolase and glutathione activation coefficient, serum magnesium, zinc and ferritin, blood hemoglobin concentration, and mean corpuscular volume was similar in both groups throughout the entire study. E-aspartate aminotransferase activation coefficient (pyridoxal status) increased from 2.02 ± 0.06 to 2.11 ± 0.06 in exercise and decreased from 2.23 ± 0.08 to 2.08 ± 0.06 in the control group (p = 0.04). E-Mg was stable in exercise (6.82 ± 0.13 to 6.85 ± 0.12 µmol/g Hb), but decreased from 7.13 ± 0.15 to 6.86 ± 0.15 in control group (p = 0.004). E-Zn increased from 0.57 ± 0.001 to 0.62 ± 0.001 in exercise subjects, but remained stable in controls (p < 0.0001).

In conclusion, a 24-week progressive fitness-type exercise program, effective enough to increase VO_2 max by 10% in female young adults, did not result in negative changes in indicators of thiamin, riboflavin, magnesium, iron and zinc status. Number of iron-deficient subjects did not increase, even when the participants' iron stores initially were small. The physiological relevance of slightly impaired E-ASTAC values following the exercise program remains to be answered.

Gradual and Rapid Weight Reduction in Weight-Class Athletes: Effects on Nutritional Status and Performance

Mikael Fogelholm[a], Risto Kostkinen[b], Juha Laakso[c], Tuomo Rankinen[d], Inkeri Ruokonen[c]

[a] Department of Nutrition, University of Helsinki; [b] Department of Physical Education, University of Jyväskylä; [c] MILA Laboratory Ltd., Helsinki; [d] Department of Nutrition, University of Kuopio, Finland

We studied the effects of two weight reduction procedures on nutrient intake, micronutrient status and physical performance in 7 male wrestlers and 3 judo athletes, aged 17–31 years. In the 'gradual' procedure (GP), a 5.0 ± 0.4% (mean ± SEM) body weight loss was achieved in 3 weeks by a weight-reducing diet. In the 'rapid' procedure (RP), 6.0 ± 0.6% of body weight was reduced in 2.4 days by fluid and diet restriction and forced sweating, followed by a 5-hour rehydration period.

Nutrient intake was calculated from 4-day food records. Compared with baseline (weight maintenance), energy intake was significantly ($p < 0.01$) lower during both GP and RP. Protein intake at baseline was 107 ± 17, during GP 71 ± 16 ($p < 0.01$ vs. baseline) and during RP 56 ± 17 g/day. Carbohydrate intakes were 302 ± 48 (baseline), 239 ± 56 (GP1) and 182 ± 55 g/day (RP, $p < 0.01$ vs. baseline). Excluding vitamin C (both procedures) and magnesium (GP), micronutrient intake during weight reductions was lower ($p < 0.01$) than during baseline. Mean thiamin, magnesium and zinc intakes were at or below the respective recommendation during weight reductions. Micronutrient status (vitamins B_1, B_2 and B_6, potassium, magnesium, iron and zinc) was assessed from blood chemistry. Most values remained stable during both GP and RP. Changes in E-ASTAC (vitamin B_6) and S-magnesium concentration were different ($p < 0.01$) between the procedures, suggesting negative trends during GP.

Sprint performance (30 m time) was similar throughout the study. Following GP, vertical jump height with extra load increased slightly (6–8%). Nevertheless, jumping results did not change significantly during RP. To assess both anaerobic performance and recovery from anaerobic load, two Wingate tests were carried out with a 30-min rest interval. We found no drastic changes, although, after GP, mean power during the last 30 s decreased slightly. Unchanged total work output suggests that anaerobic performance was not affected by either procedure.

We conclude that experienced athletes may reduce their body weight 5% with either gradual or rapid method, without impairment in sprint, power or anaerobic performance.

Influence of Nutritional Status on the Functional Capability of Spanish Elderly

R.M. Ortega[a], P. Andres[a], A. Melendez[b], E. Turrero[a], M.J. Gaspar[a], M. Gonzalez-Gross[a], G. Garrido[b], M. Chamorro[c], E. Diaz-Albo[d], O. Moreiras-Varela[a]

[a] Departamento de Nutrición, Faculdad de Farmacia, Universitad Complutense de Madrid; [b] Departamento de Fisiología del Ejercicio, Instituto Nacional de Educación Física; [c] Centro Nacionál de Medicina Deportiva (ICEF y D), y [d] Centros Gerontológicos de Caritas de Madrid, Spain

The present study analyzes the influence of nutritional status on the functional capability of 11 institutionalized elderly (4 men and 7 women) (mean age 78.1 ± 4.1 years) living in the Comunidad Autonoma de Madrid. Nutritional status was evaluated by dietetic, anthropometric, hematological and biochemical data; and the functional status was evaluated considering adiposity, strength in hands and legs bent and stretched, and flexibility. The caloric intake was inferior to the recommended amount in 90.9% of the cases, a condition leading to a deficient intake of micronutrients in an important percentage of the population. Although the caloric intake is deficient in most of the cases, 54.6%

of the elderly are overweight and 27.3% are obese, probably because of the little physical activity that we observed in this community. The hematological and biochemical data show a 27% deficiency of ferritin and serum iron. 18% have thiamin deficiency, 14% riboflavin deficiency, 25% retinol deficiency and all have a slight deficiency of folic acid and vitamin C. The most important nutritional problems that contribute to functional wastage are obesity, hypercholesterolemia and protein and micronutrient deficiency. The adverse influence of obesity and hypercholesterolemia on the functional capacity of the elderly is shown by the inverse relationship between flexibility and strength in hands and legs with degree of adiposity, thickness of skin folds and hypercholesterolemia. In reference to the diet's influence, there are positive correlations between food intake and most of the nutrients with hand and leg strength, and there are statistical significances for proteins, iron, zinc, magnesium and pyridoxine, and also for vitamin C, niacin, thiamin, folic acid and vitamin E. For blood values, the major correlation exists between functional parameters and iron, ferritin and vitamin C levels. Our results confirm the influence of nutrition on the functional capacity of the elderly and provide evidence for the necessity of improving the elderly's diet to prevent micronutrient deficiency, and also the necessity of increasing their physical activity. Both measures will promote health and the functional improvement of the elderly.

This study was supported by the Fondo de Investigaciones Sanitarias de la Seguridad Social (FISss).

Influence of the Iron Status on the Attention and Intellectual Capability of Spanish Adolescents

R.M. Ortega [a], *M. Gonzalez-Fernandez* [a], *L. Paz* [b], *P. Andres* [c], *L.M. Jimenez* [d], *M. Gonzalez-Gross* [a], *A.M. Requejo* [a], *M.J. Gaspar* [a]

[a] Departamento de Nutrición, Faculdad de Farmacia, Universidad Complutense de Madrid; [b] Departamento de Personalidad, Modificación de Conducta, Faculdad de Psiología, UNED; [c] Departamento de Nutrición y Bromatología II, Laboratorio de Técnicas Instrumentales, Faculdad de Farmacia, Universitad Complutense de Madrid; [d] Asociacion Pedagógica para le Calidad de la Ensenanza (APICE), Madrid, Spain

Dietetic, hematologic, and biochemical data were used to assess the iron status of a group of 64 adolescents (37 males and 27 females), aged 15–18 (mean age 15.94 $\pm$ 0.76 years), who study in a High School in the Comunidad Autonoma de Madrid. All were asked to keep a dietary record during 5 days, one of which had to be a Sunday. Iron intake was estimated using the Food Composition Tables of the Instituto de Nutricion (1990). The hematologic survey determined hemoglobin, hematocrit, mean corpuscular volume (MCV), mean corpuscular hemoglobin (MCH), mean corpuscular hemoglobin concentration (MCHC), serum iron and serum ferritin. These data were correlated with the scores obtained in the attention and school capability test (AT), that gave information about the verbal (V), reasoning (R) and calculus (C) capabilities.

There is a positive correlation between MCV ($r = 0.2705$), MCH ($r = 0.3370$), ferritin ($r = 0.3134$) and attention. MCV ($r = 0.2995$), MCH ($r = 0.3998$), MCHC ($r = 0.3134$) and ferritin (r = 0.3970) were also correlated with the speed capability shown on the attention test and the hemoglobin level was correlated with the calculus capability (r = 0.2905). The students who obtained higher scores in the school capability test also had better blood parameter values. This was statistically significant for serum ferritin in male students and for hemoglobin and MCHC in female students. 19.6% of the adolescents had ferritin levels lower than 12 ng/ml. Their intelligence test scores were lower than those who had serum ferritin ≥ 12 ng/ml. This difference is statistically significant, in both sexes, for the verbal intelligence and the school capability test scores.

Our results suggest that there is an influence of the iron status on the attention and school cognitive functions in the studied adolescent group.

This study was supported by Precompetitive Groups of the Universidad Complutense de Madrid.

Dietary Patterns of a Group of Football Players in Madrid (Spain)

M. Gonzalez-Gross[a], *R.M. Ortega*[a], *P. Andres*[b], *A.M. Requejo*[a], *M.A. Herrador*[c]

[a] Departamento de Nutrición, Faculdad de Farmacia, Universidad Complutense de Madrid; [b] Departamento de Nutrición y Bromatología II, Laboratorio de Técnicas Instrumentales, Faculdad de Farmacia, Universidad Complutense de Madrid; [c] Servicios Medicos del Real Madrid, Spain

Dietary frequency intake habits were evaluated in a group of 41 male National League football players, aged 17–21, by a self-completed 'dietary frequency intake' questionnaire. All of them used olive oil daily to cook their foods. 7.3% had a daily intake of nuts, 24% consumed them 5 times a week and 53.6% once or twice per week. 63.4% never ate entrails and 17% consumed them once or twice per week. This contrasts with the high meat, sausages and jam intake, because 29% of the population ate meat daily and the rest 5 times a week; for sausages there was 44% daily intake, 41% several days a week and 12% once or twice a week. On the other hand, only 1 player ate fish daily, 36.5% ate fish 5 times a week, 46% ate fish once or twice a week, 10% ate fish once a month and 2 never ate it. 80.5% drank milk every day, 12% several times per week and 5% once or twice per week. 51% ate cheese once or twice a week, 29% several times per week, 5% every day and 13% never ate cheese. Eggs were consumed by 12% on a daily basis, by 63% several times a week and by 24% once or twice per week.

In reference to the carbohydrate sources, 19.5% ate potatoes daily, 71% ate them several days per week and 7% once or twice per week. But when we asked how they cooked the potatoes, they all answered fried. Noodles were consumed daily by 10% of the population, several times per week by 54% and once or twice per week by 37%. Rice was eaten daily by one player, 29% ate rice several times per week, 66% once or twice per week and 1 player ate rice once a month.

Only 2 players ate vegetables daily, 41% ate vegetables 3–5 times per week, 49% once or twice per week and 1 of the players never ate vegetables. 66% of the football players ate fresh fruits daily, 32% ate fruits several times per week and one consumed fruits once or twice per week. Legumes were consumed one or twice a week by 46% and 3–5 times per week by 42%.

Metabolic Effects of Everyday Exercise Until Exhaustion

Malgorzata Zendzian-Piotrowska, Jan Gorski
Department of Physiology, Medical School, Bialystok, Poland

There is a vast literature on metabolic adaptation to endurance training. The training programs used usually consisted of nonexhaustive exercise bouts of different duration, 5 days a week, for several weeks. In the following study rats were exercised until exhaustion everyday, 7 days a week, 28 days, on a treadmill set at $+10°$ incline and moving at a speed of 1,200 m/h (T group). The control rats (C) were housed in the same conditions as T but did not run. All rats had unrestricted access to food (a rodent's pellet diet) and water. The rats were weighed every morning. Also an amount of food consumed every day by each rat was weighed. C rats increased their body weight from 251 ± 17 g on the first day of the experiment to 299 ± 31 g on the 28th day. On the contrary, the T rats reduced their body weight from 257 ± 21 g on the first day to 221 ± 20 g on the 28th day. The amount of food consumed in the C group was 22.6 ± 4.3 g and 25.7 ± 3 g on the 1st and 28th days, respectively. In the T group, there was a reduction of the food consumed during the first 3 days. Then, it reached the preexercise value on the fourth day and 30.2 ± 3.7 g on the 14th day and stabilized thereafter. Time of running until exhaustion increased from 177.9 ± 29.3 min on the first day to 327.8 ± 27.8 min on the 28th day. In each group different biochemical variables were determined at rest and after exercise until exhaustion. After exercise until exhaustion the glycogen concentration in skeletal muscles (white and red vastus, soleus, diaphragm) and in the heart of T rats was about two times higher whereas in the liver several times higher than in C rats. The T rats did not develop hypoglycemia after exercise until exhaustion. The postexercise plasma free fatty acid concentration in the T group was much lower than in the C group whereas the concentration of urea, creatinine and uric acid was similar in both groups. 24-hour urine nitrogen excretion after exhaustion exercise in the T group was higher than in the C group.

It is concluded that everyday exercise until exhaustion is well tolerated. However, the animals gradually lose their body weight. In the process of training of this type very efficient mechanisms develop conserving the body's carbohydrate reserves but not protein reserves.

Protein Satiety Comparing Three Meat Species

J. Hamilton, D. Oberleas, M. Harden
Food and Nutrition Program, Texas Technical University, Lubbock, Tex., USA

This study involved 26 Texas Tech female and 15 male students of normal weight ranges, recruited after screening for taste preferences and demographics. A preload of beef, chicken and fish were fed to subjects followed by fruit-flavored yogurt shakes as the test meal. Subjects were seated in sensory panel booths where odor, lighting and conversation were controlled. The preload levels of protein sources were determined by offering unlimited quantities of beef, chicken or fish to the subjects on 3 successive test days. Preloads of 120 g were selected for females and 200 g for males. Subjects were allowed 15 min to consume the protein preload then at time parameters of 20 or 40 min after the consumption of the preload, yogurt milkshake was ingested. Subjects were asked to consume 100% of the protein source and at the specified time were given the milkshake to drink until satiety (satisfaction or a feeling of fullness) was attained. Quantitative measurements of the milkshake consumption were used to determine the satiety value of the protein source. The data were treated with a 3-way ANOVA with replications. The four variables were protein source, lean beef, chicken, and fish; time delay; particle size; and gender. There was a significant effect for protein source (p < 0.0001) with the mean ± SEM as: beef 290.1 ± 20.9, chicken 348.3 ± 20.1, and fish 361.1 ± 19.0 g, and a difference between gender (p = 0.026) with females consuming 302.7 ± 12.2 and males 386.0 ± 23.1 g shake. Present results indicate that beef satiates longer than chicken or fish at the same fat level.

Nutritional Assessment of a Team of Basketball Players

M. Hassapidou, K. Kleftouri, E. Efstratiou, S. Kitsou
Technological Educational Institution of Thessaloniki, Greece

The purpose of this study was to assess the nutritional status of the elite athletes of a basketball team. Body weight, height and skinfold thickness were measured and BMI was calculated for all 9 athletes. Dietary recall records were taken before and during the period of training (June and September, respectively). Weight records were taken before competition (end of October). All dietary records were analyzed for energy and nutrients. The mean energy intake of the athletes was 4,523 kcal before training, 3,011 kcal during the period of training and 4,596 kcal before competition. Their mean protein intake was 124 g (11%E) before training, 122 g (16%E) during training and 180 g (16%E) before competition. Their mean fat intake was 227 g (45%E) before training, 136 g (41%E) during training and 155 g (31%E) before competition. Their mean carbohydrate intake was 531 g (44%E) before training, 347 g (43%E) during training and 662 g (53%E) before competition. All results showed that dietary intakes varied among the athletes but in general they reflected unbalanced nutrition. The diet was more balanced before competition.

Total serum cholesterol, LDL-chol, HDL-chol, triglycerides, albumin and total proteins were analyzed before the period of training. Two athletes had high levels of cholesterol, high levels of LDL-chol and low levels of HDL-chol, which correlated well with their unbalanced nutrition. The anthropometric measurements showed that 4 athletes had BMI greater than 25% and 3 had total body fat greater than 12%.

Diet in Athletes (USA versus East Germany)

Lothar A.J. Heinemann[a], Ann C. Grandjean[b]
[a] Center for Epidemiology and Health Research Berlin, Zepernick, Brandenburg, FRG; [b] International Center for Sports Nutrition, Omaha, Nebr., USA

Comparative dietary studies were performed in 275 athletes in the USA and 167 elite athletes of the former German Democratic Republic. Both studies were based on a 3-day dietary protocol method using a computer program with the respective national food tables for calculating the nutrients. The aim of this comparison was to empirically back-up the hypothesis that there is no very special dietary precondition for successful performance, except a healthy diet.

The caloric intake tends to be slightly higher in East German athletes. The protein intake (adjusted to body weight) is similar in athletes and the respective population. The fat intake, however, is higher in East Germany, both in athletes and particularly in the population, compared to the USA. After adjustment for total calories consumed, the dietary behavior of athletes is closer to the general population in East Germany than in the USA. Differences between athletes are somehow reflecting the differences between the population regarding dietary behavior. The same applies for micronutrients: vitamin C intake and minerals are not very different (somewhat higher in the USA).

Conclusion: There are more similarities than differences among various sports in both populations. There are no empiric hints for a 'more successful' special diet in elite athletes of both countries. However, athletes are generally closer to the desirable long-term 'Recommended Dietary Allowances' than the respective population, i.e. prefer a more healthy diet.

LanguaL – The Universal Language for Food

Thomas C. Hendricks
US Food and Drug Administration, Washington, D.C., USA

LanguaL is an automated method for describing, capturing and retrieving data about food. It was developed by the Center for Food Safety and Applied Nutrition (CFSAN) of the Food and Drug Administration (FDA). LanguaL is based on the concept that: (a) Any food (or food product) can be systematically described by a combination of characteristics. (b) These characteristics can be categorized into viewpoints and coded for computer processing. (c) The resulting viewpoint/characteristics codes can be used to retrieve data about the food from external data bases.

Each viewpoint is called a 'factor'. There are presently 15 such factors (e.g. product type, food source, etc.). The characteristics related to them are called 'factor terms'. For example, consider the 'mixed dish' food product 'homemade lasagna'. Its Product Type factor term would be 'pasta dish'. Its Food Source factor term would be durum wheat (because of the predominate noodles), etc. Thus, the lasagna dish can be described in sufficient detail to allow retrieval of information from external data bases using factor term codes (e.g. pasta dish = A220, durum wheat = B1312), representing its various characteristics. Furthermore, a list of ingredients or a recipe can be associated with each food (food product) in order to give more detailed information, particularly about complex composite foods such as the lasagna example cited above.

A computer record contains the food (product) name and assigned factor term codes, as well as product identification codes for accessing external data bases [e.g. FDA Total Diet System, US Department of Agriculture (USDA) Handbook No. 8]. An associated computer record contains the related ingredient/recipe data. Information is retrieved by searching the foods (products) thus coded (currently approximately 25,000) using selected factor term codes and/or ingredient recipe identifiers to obtain pointers to outside data bases. Thus, a single search can result in data being obtained from a variety of sources, including analytical and consumption data bases maintained by the FDA and the USDA as well as other countries (e.g. France, Denmark).

Nutritional Evaluation of Young Israeli Athletes: Are Basic Nutritional Requirements Being Met?

N. Israel, J. Kaplanski
Clinical Pharmacology Unit, Ben Gurion University of the Negev, Beer Sheva, Israel

The purpose of this study was to evaluate the adequacy of nutritional intake of young athletes in Israel. The study included 26 female gymnasts, ages 7–13; 10 male swimmers, ages 12–18; 27 female physical education students, ages 18–21; and 18 sedentary girls, ages 8–12. The study included: collection of information concerning nutritional intake and physical activity by keeping 3-day diaries; collection of anthropometric data by measuring height, weight and skinfolds; and collection of biochemical data by blood tests. All the data was analyzed by a newly developed computerized nutritional system. The main results of this study were: (1) Both groups of athletes showed typical inadequacy in nutritional intake of several nutrients: a low intake of Ca, Zn, Cu, Mg, vitamin A, vitamin B_1, vitamin B_2 and nutritional fibers, and a very high intake of sodium. (2) Both groups showed an adequate energy intake related to daily energy expenditure. (3) No cases of underweight or acute growth retardation were found among the subjects. The gymnasts had a significantly smaller body stature compared to the sedentary girls, although weight to height ratio, and energy consumption were found to be normal. (4) Fat percent of the athletes was significantly lower than in the sedentary group.

It was concluded that the intake of nutrients in the young gymnasts and swimmers is far from being optimal, and in many cases not even the minimum required, in spite of adequate caloric intake. Although no acute signs of malnutrition were found, this pattern of intake might cause subclinical deficiencies in the future, and subsequent chronic damage to growth and athletic performance.

Aerobic Capacity, Coronary Heart Disease Risk Factors Status, and Dietary Habits in Cretan Obese Children

E. Lydaki, A. Kafatos, E. Papadakis, P. Konstantinakos, D. Kounali
University of Crete, Iraklion, Crete, Greece

Obesity has a high prevalence among Greek children. In a selective sample of overweight children from Crete the aerobic capacity, CHD risk factors and dietary habits were studied. The study population consisted of 69 overweight children having a BMI over 20.1 kg/m^2 and 66 children having BMI under 20 kg/m^2 (90th percentile of HANES study). The mean BMI was 21.9 kg/m^2 for overweight and 17.5 kg/m^2 for normal children. Food intake was recorded by a 24-hour dietary recall and a weekly frequency of food consumption recall method. Aerobic capacity test was based on Cooper's protocol. 10 ml of blood was obtained for total serum cholesterol and HLD-C levels and 5–10 mg of adipose fat obtained by needle aspiration from 40 children for fatty acid composition analysis. The overweight children reported lower daily sugar ($p < 0.02$) and energy intake ($p < 0.05$) than the nonobese. The saturated fat consumption was 12%, monounsaturated 15% and polyunsaturated 3% of total energy intake for both overweight and normal children. Obese children had higher percentage of C16:1 in adipose fat than normal. BMI had a negative correlation with the aerobic capacity (p < 0.0005) and serum HDL-C levels (p < 0.02). Ten percent of overweight and 18% of normal children had serum cholesterol over 200 mg/dl. Overweight boys had higher systolic blood pressure levels than normal (p < 0.005). The fatty acid composition of adipose tissue analysis showed no significant difference in obese and nonobese children with the exception of elevated percentage of C16:1. The higher percentage of C16:1 in overweight children suggests a higher sugar and energy consumption in the past several years. This finding does not agree with the current dietary data indicating a lower sugar and energy intake by obese children. In conclusion, overweight children have lower aerobic capacity, serum HDL-C levels and higher systolic blood pressure. Obese children tend to underreport energy and sugar intake which has been confirmed by adipose tissue composition analysis.

Physical Activity, Health and Nutritional Status of the Medical Students at the University of Crete

D. Vassilakis, A. Kafatos, D. Kounali, D. Labadarios
University of Crete, Iraklion, Crete, Greece

The health, nutritional status and physical activity of the medical students at the University of Crete were studied in three classes (1989, 1990, 1991) of 155 students, who participated in a 30-hour required course of clinical nutrition. The evaluation of health and nutritional status was part of the course intended to motivate students about their own health. The analysis of the results of each class was used as an example of community nutrition approach. The course was given in the 6th semester, the average age was 22.3 years for boys and 21.7 years for girls. The evaluation of health status included: clinical examination, venipuncture, anthropometry (the percentage of fat derived from two skinfolds), blood pressure, dietary history, and smoking habits. The physical activity of the students was evaluated by standardized questionnaire which took into account intensity of activity, duration in time and frequency per year. Caloric expenditure was calculated by the following formula: calorie expenditure = time of exercise × energy expenditure/min/kg × body weight. The results showed that BMI for boys was 23.7 and for girls 22.7. The percentages of body fat was 27.4 for boys and 28.3 for girls; 43.9% of the boys smoked while 37.5% of the girls smoked. SBP and DBP was of no significant difference between boys and girls. Mean total serum cholesterol was 189 mg/dl for boys and 185 mg/dl for girls (NS) and the HDL was 39 mg/dl for boys and 52 mg/dl for girls (p < 0.001). Serum triglyceride was 87.3 mg/dl for boys and 62.2 mg/dl for girls (p < 0.001). The LDL/HDL ratios were 5.3 for boys and 3.4 for girls (p < 0.001). Only 16.6% of the boys and 18% of the girls were involved in regular physical activity. The medium weekly energy expenditure was 192 kcal for boys and 197 kcal for girls. Only 3 girls and 1 boy had more than 4,000 kcal/week energy expenditure. The medium daily energy intake was 2,603 kcal for boys and 1,672 kcal for girls. The total fat intake was 38.5% of the energy intake for boys and 39.3% for girls. The saturated fat intake was 12.7% at total energy intake for boys and 13.5% for girls. The analysis of covariance showed that the lack of physical activity had a significant negative effect on total serum cholesterol, triglyceride and the LDL/HDL ratio, when controlling for total body fat, caloric and saturated fat intake and W/H circumference. In conclusion, approximately 75% of the medical students at the University of Crete had very limited physical activity (weekly energy expenditure less than 500 kcal) while the prevalence of smoking was high. The analysis of covariance indicated that physical activity can significantly improve the lipoprotein profile of the students.

The Role of Olive Oil in Healthy Nutrition and Fitness

A. Kiritsakis, M. Hassapidou
School of Food Technology and Nutrition, TEI, Thessaloniki, Greece

Olive oil has been one of the most important products in the Mediterranean countries, where its health value has been acknowledged since ancient times. Olive oil has a positive role in the treatment of hyperchlorhydria, reduces gallstone formation and has potential value in the nutritional therapy of diabetes. It also has a radioprotective effect on the skin. Olive oil has positive effects on growth and development and possibly on aging. It is absorbed more rapidly and efficiently compared to other oils. A number of epidemiological studies such as the Seven Countries Study have shown that populations, like the Greek population, with a high consumption of olive oil have lower incidence of coronary heart disease. Olive oil seems to reduce levels of LDL while it either has no effect or increases the levels of HDL. It has also been shown that olive oil has a positive effect on leukotriene formation and reduces platelet aggregation. The recognized nutritional and biological value of olive oil is related to its chemical composition. It is considered a high monounsaturated oil with an appropriate content of the essential polyunsaturated fatty acids for humans, a high content of sterols, total phenols and other constituents which increase its stability. When olive oil is heated a lower content of trans fatty acids (TFA) is formed compared to other oils and fats. Since successful athletic performance is based on a healthy and well-balanced nutrition, olive oil, better than any other fat, can provide a basis in the planning of a healthy diet for athletes.

Micronutrient Intake and Mortality of Coronary Heart Disease

Paul Knekt, Antii Reunanen, Ritva Jarvinen, Ritva Seppanen, Arpo Aromaa
The Social Insurance Institution, Helsinki, Finland

The relationship between the intake of retinoids, carotenoids, vitamin C and vitamin E and the subsequent risk of coronary heart disease was studied among 5,133 men and women aged 30–69 years and initially without known heart disease in a cohort study in Finland. Food consumption was estimated by the dietary history method covering the total habitual diet during the previous year. Altogether 244 new fatal coronary heart disease cases occurred during a mean follow-up of 12 years. An inverse association was observed between vitamin E intake and coronary heart disease mortality both among men and women with a relative risk of 0.68 (p for trend = 0.01) and 0.35 (p < 0.01) between the highest and lowest tertiles of the intake, respectively. A similar association was observed for intakes of vitamin C and carotenoids among women and for intakes of vegetables and fruits among men and women. The associations were not due to confounding by age, smoking, serum cholesterol, hypertension, body mass index or energy intake. The results are in accordance with the hypothesis that antioxidant vitamins may protect against coronary heart disease. It cannot, however, be excluded that food sources rich in these micronutrients also may have other constituents which may provide the protection.

Augmented Blood Pressure during Stress Test Measurement in Youths with High LDL-Cholesterol

M.E. Makris, I.G. Kazoulakis, P.A. Beloussis
Cardiodiagnostic Center, Rhodes, Greece

The purpose of this study is the observation of blood pressure (BP) levels during tread mill stress tests (ST) of youths with high LDL-cholesterol (LDL-CH) (160 mg/dl). Ten young athletes (group A), aged 14–16 years, with LDL-CH ± 160 mg/dl were studied. The results were compared to those of 12 young athletes aged 14–16 years, having normal cholesterol levels, when given a ST (group B). The ST was given based on the Bruce Protocol and 95% of the predicted maximum heart rate was achieved. Simultaneously, blood pressure was measured at rest, during the ST peak and 10 min after the test. Both groups, at rest, had normal BP rates, along with the same average age and similar height and weight.

The results are as follows:

		Group A (n = 10)	Group B (n = 12)	p
At rest	SBP	115 ± 15	112 ± 13	MS
	DBP	70 ± 7	60 ± 8	±01
Stress peak	SBP	182 ± 20	160 ± 25	±04
	DBP	75 ± 10	70 ± 8	±04
10 min after test	SBP	120 ± 10	110 ± 9	±01
	DBP	64 ± 5	60 ± 8	MS

Conclusions. (1) Youths with high LDL-CH have higher blood pressure rates during a stress test. (2) A stress test can offer important information during examinations of youths with high LDL-CH.

Physical Activity and Nutritional Status in Anorexia nervosa

A. Marcos, P. Varela
Instituto de Nutricion CSIC, Facultad de Farmacia,
Universidad Complutense de Madrid, Spain

In industrialized nations thinness is becoming more interesting as a symbol of fitness, a fact which may be linked to illness associated with altered food behavior, affecting especially certain groups of the young female population. These reasons might explain a higher prevalence of syndromes such as anorexia nervosa (AN) in elite sportswomen (dancers, gymnasts, etc.). Therefore, the goal of this work was to study the nutritional status throughout immunocompetence evaluation in young women with different physical activity levels and degree of suffering from eating disorders. Leukocyte counts were higher while lymphocyte counts remained unmodified due to a lower lymphocyte percentage found in the AN group which practiced sports in comparison with the nontraining group. Cell-mediated immune function determined by delayed hypersensitivity skin test was shown to be depleted in the AN group who trained with physical exercise, since the 'score' (the whole response to the seven antigens administered) was 30% lower than the value found for the sedentary AN group, although no modifications were observed in the number of positive response to the antigens. In relation to innate immunity, total hemolytic serum capacity increased 32% in AN group practicing sports, in spite of the fact that both C3 and C4 complement factors were not altered. Previously, we had pointed out similar outcome in healthy young women training with slight physical activity. Therefore, nutritional status determined by immune system evaluation appeared to be modulated by physical exercise in the anorexic young women tested in this study.

Hemolytic and Biochemical Alterations in Male Athletes: A Call for Improved Nutrition

O.O. Odedeyi
General and Exercise Physiology Research Laboratory, Ogun State University,
Ago-Iwoye, Nigeria

The study was embarked upon to ascertain the outcome of a 12-week aerobic exercise training program on the hemolytic and biochemical status of red blood cells. Sports anemia has been observed as a transient phenomenon in athletes of all grades, age and activities. Although transient in nature, it could develop into frank and clinical anemia in already anemic athletes. The present study, however, considered the pre- and post-training effects of aerobic (jogging) training program on the osmotic fragility of the red cells and the rate of conjugated, unconjugated and total bilirubin in athletes at the two different levels. The pre-test, post-test control group experimental design was used for the study. Hypotheses were tested at the 0.05 alpha level with critical value at 3.59. The mean, standard deviation and ranges were used as descriptive statistics. One was analysis of variance (ANOVA) used at pre-training to test groups homogeneity and analysis of covariance (ANCOVA) was applied to the pre-test and post-test data to determine the effects of training on the parameters. Osmotic fragility graphs were plotted for pre- and post-training percent hemolysis.

The post-test results showed significant difference $p < 0.043$ for osmotic fragility; however, the increase in conjugated, unconjugated and total bilirubin was not statistically different from control subjects. The cumulative and derivative fragiligram for the post-test result depicted increased hemolysis which culminated in sports anemia. The results of this investigation unequivocally pointed to an incipient sports anemia which requires improved nutrition for athletes to supplement body activities in the biochemical process of hematopoiesis.

The Tripod of Malnutrition, Bioenergetics and Fitness

J.O. Olowookere
Biomembranes and Bioenergetics Research Laboratory, Ogun State University,
Ago-Iwoye, Nigeria

Overnutrition culminating in obesity, and undernutrition resulting in kwashiorkor or marasmus are key malnutritional diseases with divergent etiologies, yet showing similarities of bioenergetics problems. Our present investigation on the hepatic mitochondrial respiration and ATP synthesis in induced kwashiorkor and obesity syndromes is to experimentally demonstrate this triangular relationship existing among malnutrition, bioenergetics (mitochondrial energy transduction processes in eukaryotic cells) and physical fitness.

Kwashiorkor was simulated by feeding a low-protein/high-calorie diet (LPD) to weanling rats. Obesity was induced by feeding a high-protein/high-calorie diet (OBD). Control rats were fed with a commercially produced diet (CPD). The mineral salts and vitamins of the three dietary regimens were the same. Duration of feeding was 21 days. Our results on the mitochondrial respiration in the presence of malate or succinate (normal respiratory substrates in man) revealed that the resting respiration (state 4) was 23.9% higher in kwashiorkor and 29.1% higher in dietary obese rats. However, the active respiration (state 3) was 34.8% lower in kwashiorkor and 43.3% lower in obese rats when compared with controls. State 3 respiration is biochemically linked with ATP (energy) production. The ratio of state 3 to state 4 gives the respiratory control ratio (RCR). The respiratory control ratios (indices of energy production) were 51.1% in kwashiorkor and 43.8% in obese rats relative to the values obtained in control rats.

It is concluded from our studies that kwashiorkor, and obesity lead to defective energy transduction processes by interfering with mitochondrial bioenergetics. The above, invariably, negatively affects energy expenditure, work capacity and overall physical fitness.

Nutrient Intake of Greek Elite Athletes

M. Papadokostantaki, K. Pavlou, M. Hassapidou
Department of Exercise Physiology and Nutrition, Hellenic Sports Research
Institute, Olympic Athletic Center of Athens, and Nutrition Department of
Technological Educational Institution of Thessaloniki, Greece

Food intake was measured in 41 elite athletes of the Greek National Teams of rowing (R), volleyball (V), cycling (Cy) and weight lifting (WL). The study took place during competition Cy and V, and training R and WL. The athletes recorded their daily food and supplement intake in a specially designed diary for 5 days. They were instructed on how to fill in the diaries which were checked afterwards. Dietary records were analyzed for energy macro- and micronutrients. Mean daily nutrient intake is shown in table 1.

Table 1. Mean daily nutrient intake

Nutrient intake	Rowing (n = 16)	Volleyball (n = 14)	Cycling (n = 6)	Weight lifting (n = 5)
kcal	4,211 ± 227	3,500 ± 338	5,023 ± 780	3,157 ± 420
Protein (P),g	188 ± 19 (19%)	168 ± 16 (16%)	185 ± 12 (15%)	156 ± 28 (20%)
Carbohydrates (C), g	587 ± 15 (55%)	460 ± 26 (56%)	633 ± 98 (50%)	289 ± 128 (37%)
Fat (F), g	124 ± 22 (26%)	111 ± 19 (28%)	189 ± 20 (34%)	153 ± 93 (44%)

The average carbohydrate and protein consumption of R and V met the current Recommended Dietary Allowance (RDA) (P = 15–20%; C = 50–55%; F = 25–30%). Carbohydrate consumption was marginal while fat intake was higher than the RDA in Cy. The weight lifters' dietary intakes reflected unbalanced nutrition. Their mean fat intake (44%) was higher and carbohydrate intake (37%) lower than the RDA. Micronutrient intake in food and supplement combined exceeded the RDA values for all groups examined.

Effect of Systematic Administration of a Complex Vitamin/Nutritional Supplement on Physical Work Capacity

P. Pshendin, V. Zagranzev, V. Rogozkin
Research Institute of Physical Culture, St. Petersburg, Russia

The possibility of the development of athletes' physical work capacity anaerobic mechanisms by a systematic administration of a complex vitamin/nutrition supplement 'Mixovit-Forte' has been investigated in the given work. 21 volunteers of top class qualification took part in the investigations (athletes Alpine combination). The average age was 15.2 ± 1.1 years; the height was 168.0 ± 3.2 cm; body weight was 58.0 ± 2.6 kg. The experimental group (n = 10) received 75 g each of 'Mixovit-Forte' (200 kcal) in the form of water cocktail during the period of their recovery after all physical exercise. The control group (n = 11) received placebo for 12 days. The product consisted of a mixture of milk proteins produced by means of the technology of membraneless osmosis, 14 basic vitamins, 13 macro- and microelements, organic acids, and flavoring supplements. The athletes received the structure of recommended standards. The level of maximal physical work capacity was determined by anaerobic productivity in short-term exercises of maximal power. The alactate power (ALP) of work was judged according to maximal value of mechanical productivity on a bicycle ergometer for 6 s, then 5–10 min of rest later the alactate-glycolytic power (ALGP) was determined – the work during 20 s. The biochemical criteria of alactate-glycolytic working power and capacity was determined according to the content of creatine (Cr) and inorganic phosphate (P_i) before and after the test loading on a bicycle ergometer. The investigation was carried out before and after the 12-day administration of the product in the morning, after a standard breakfast.

It was shown that under the influence of training, both ALP and ALGP increased from 11.6 ± 0.3 to 12.3 ± 0.3 w/kg and from 7.6 ± 0.3 to 8.6 ± 0.3 w/kg, respectively. In this case delta Cr and delta P_i were 57 ± 8 and 0.19 ± 0.03 mmol/l. However, the administration of 'Mixovit-Forte' led to a greater and more reliable alteration of physiological and biochemical indices of work capacity. So, the specific value of ALP changed from 11.5 ± 0.4 to 13.3 ± 0.4 w/kg, while delta Cr was 83 ± 10 mmol/l and delta P_i 0.30 ± 0.03 mmol/l.

Thus, one may suppose that the systematic administration of a product with high alimentary density under the conditions of training and habitual nutrition creates an advantageous background for the formation of sports work capacity at the expense of increasing anaerobic power.

Effects of Combined Diet and Exercise Intervention on Serum Lipoprotein Cholesterol with Special Reference to Apolipoprotein E Polymorphism

Tuomo Rankinen, Rainer Rauramaa, Sari Vaisanen, Esa Hämäläinen,
Ilkka Penttila, Pekka Oja, Ilkka Vuori
Kuopio Research Institute of Exercise Medicine, Kuopio;
UKK Institute, Tampere, Finland

The aim of this study was to evaluate the combined effects of fat reduced diet and moderate aerobic exercise on serum lipoprotein cholesterol. 34 sedentary men aged 51–53 years were randomly assigned to a reference and 36 to a diet plus exercise (DEXE3) group. During the 6 month's intervention subjects took moderate aerobic exercise three times a week and followed moderate fat (34.5% of energy, E%) diet with saturated fatty acids contributing 12.7, monoenes 13.5 and polyenes 6.0 E%. Cholesterol intake was 256 mg/day. In the DEXE3 group LDL decreased by 0.21 mmol/l, HDL and HDL_2 increased by 9.2 and 23.0%, respectively. Respective changes in the reference group were 0.02 mmol/l, 4.8 and 12%. In the DEXE3 group men not carrying the E4 allele in the apo-E gene (n = 28) showed much greater changes in all lipid parameters than those provided with that allele (n = 8) (−0.27 vs. −0.03 in LDL, 0.13 vs. 0.03 in HDL and 0.18 vs. 0.08 in HDL_2). In the reference group subjects without E4 allele showed 0.06 mmol/l increase whereas men carrying that allele had an equal decrease in LDL cholesterol. Increase in HDL was greater in men not carrying the E4 allele (0.08 vs. 0.03) but HDL_2 was elevated similarly in both phenotypes in the referents. Thus, these results suggest that combination of moderate fat, low cholesterol diet and mild to moderate aerobic exercise favorably influence serum lipoprotein concentration in sedentary middle-aged men, but the effects may be more profound in men without the E4 allele in their apo-E gene.

Evaluation of the Fat-Diet Composition of a Spanish Male Military Population with High Physical Activity

F.J. Sanchez-Muniz[a], *C. Cuesta*[a], *A. Domingo*[b]

[a] Instituto de Nutricion y Bromatologia (CSIC-UCM), Facultad de Farmacia, Universidad Complutense de Madrid, y

[b] Grupo de Estudios Cardiovasculares de las Fuerzas Armadas (Spain), Servicio de Medicina Interna, Hospital Militar Generalisimo Franco, Madrid, Spain

Dietary components as well as dietary habits have been implicated in major public health diseases, dietary fat and saturated fatty acid (SFA) content being especially important in the pathogenesis of coronary heart disease (CHD).

The aim of this work is to study both the dietary intake and dietary fat composition of a military male population with high physical activity in the Madrid area, Spain. A quarter of this area was randomly selected. Subjects whose daily ethanol intake was over 40 g were excluded from the study. The diet was recorded and studied over a period of 12 days in 391 men aged 20–65 years. The diet average composition was calculated per head and day, using food composition tables for the raw food weights, and the hypercholesterolemic-atherogenic potential of the average diet using the equation of Connor et al. (CSFI) and the Keys et al. index (KAGI). The polyunsaturated/saturated fatty acid ratio (P/S) was also obtained.

The detailed analysis of the diet eaten by this group of people (very homogeneous group regarding occupation and life style) shows that it was the Mediterranean dietary pattern described by several authors. Fat intake (128.0 ± 23.6 g) represents 34% of total energy intake, saturated fatty acids (SFA), monounsaturated (MUFA) and polyunsaturated (PUFA) being 9.9, 12.7% and 8.8% of the total amount of kcal, respectively. This fat pattern is considered appropriate in maintaining low the low-density-lipoprotein cholesterol levels which could take place after eating diets with a high PUFA content. Daily consumption of fish and fish products was 72.5 g, fat and PUFA from fish being 1.99 and 0.32 g, respectively. The P/S ratio was 0.81, and the CSFI of the average diet 19.2 per 1,000 kcal. According to Connor et al., a 1,000-kcal western diet has an average CSFI of 24.6, but recommends a CSFI between 8.2 and 17.5 for CHD prevention. The KAG index was 14.1, lower than the one found for the Spanish diet in 1985. The cholesterol consumption was 563.4 ± 275.0 mg/day, higher than that recommended by various experts. The diet of this military population resembles the diet in Spain of 10 years ago with rather well-balanced energy coming from fat and also with an appropriate fatty acid profile for prevention of CHD.

Profiles of Omega–3 Fatty Acids and Antioxidants in Common Purslane throughout Plant Development

Helen A. Norman[a], *Artemis P. Simopoulos*[b]

[a] USDA-ARS, Beltsville, Md.;

[b] The Center for Genetics, Nutrition and Health, Washington, D.C., USA

We have previously shown that common purslane *(Portulaca oleracea)* is the highest source of (18:3ω3) alpha-linolenic acid [1] relative to both green leafy vegetables common in the European and American diet and other weed species currently evaluated. One hundred grams of purslane (one serving) contains about 300–400 mg of 18:3ω3, 12.2 mg of alpha-tocopherol, 26.6 mg of ascorbate; 1.9 mg of beta-carotene; and 14.8 mg of glutathione. Additional reports have also identified purslane as the only higher plant known to produce eicosapentaenoic acid (20:5ω3), docosapentaenoic acid (22:5ω3) and docosahexaenoic acid (22:6ω3). In order to complement these studies, a complete profile of leaf fatty acids was prepared throughout development of wild purslane grown from seed in a controlled environmental chamber. Simultaneous analyses of alpha-tocopherol and other antioxidants were also conducted. Levels of linoleic acid (18:2) and linolenic acid (18:3) increased during plant development, and there were simultaneous increases in alpha-tocopherol and reduced glutathione for the first 45 days under the growth conditions imposed. Levels of polyunsaturated fatty acids, total membrane lipids, and antioxidants subsequently declined. Seeds of cultivated purslane obtained from different sources in the USA or Europe were grown in the same experimental conditions. It was confirmed that the relatively high levels of omega–3 fatty acids and alpha-tocopherol present in wild plants were maintained in the leaf tissue of cultivated purslane. Purslane, either in the wild or cultivated form, is a good source of omega–3 fatty acids and antioxidants and can be an important vegetable for human consumption.

Reference

1   Simopoulos AP, Salem N Jr: Purslane: A terrestrial source of omega-3 fatty acids. N Engl J Med 1986;315:833.

Differences in the Fat-Free Body to Height Relationship among Young, Older, and Mature Adults

M.H. Slaughter, C.B. Christ, R.A. Boileau, R.J. Stillman
University of Illinois, Urbana, Ill., USA

Coincident with advancing age is a decline in fat-free body (FFB) and stature (HT). The purpose of this study was to determine whether the relative musculoskeletal size differs among age groups and between genders. The subjects (n = 220) were categorized by age into three groups: young, (YA, 20–29 years), older (OA, 50–59 years), and mature (MA, 60–69 years) adults. Percent fat was estimated from a multicomponent equation which accounted for the variability in body density and the mineral and water content of the FFB. FFB was derived by multiplying the fraction of fat by body weight and subtracting the product from body weight. Least squares regressions of the log of FFB on the log of HT yielded the following within group FFB to HT relationships:

Male	YA:	$\log FFB = 1.819 \, (\log HT) - 2.28$; SEE = 0.04
Male	OA:	$\log FFB = 2.092 \, (\log HT) - 2.91$; SEE = 0.05
Male	MA:	$\log FFB = 3.018 \, (\log HT) - 5.01$; SEE = 0.03
Female	YA:	$\log FFB = 1.649 \, (\log HT) - 1.99$; SEE = 0.04
Female	OA:	$\log FFB = 1.661 \, (\log HT) - 2.05$; SEE = 0.04
Female	MA:	$\log FFB = 1.171 \, (\log HT) - 0.99$; SEE = 0.04

Significant (p < 0.05) gender differences were observed within each age group, with males having a greater amount of FFB per unit of HT (20.0–21.6 kg) than their female counterparts, depending upon the particular age group considered. Within the male sample, the YA group had significantly greater (4.2 kg) FFB per unit of HT than the MA group. In contrast, within the female sample, the relative musculoskeletal size was significantly greater in the YA group when compared to the OA (3.4 kg) and MA (5.8 kg) groups. Hence, it appears as if relative musculoskeletal size declines with increased age in both genders, with the drop off occurring earlier in the females than in the males.

Glucose Changes of Pregnant Women in Response to Prolonged Submaximal Exercise

H.N. Soultanakis, R. Artal
University of Southern California, Los Angeles County Hospital,
Los Angeles, Calif., USA

The nutritional requirements of pregnancy affect glucose homeostasis when the mother does not consume food at regular intervals. If pregnancy is already considered a stress upon the maternal glucose metabolism, then prolonged moderate intensity exercise may be considered an additional stress. This study was designed to compare the extent and the rate at which blood glucose levels fall in pregnant (25–35 weeks of gestation), versus that which occurs in nonpregnant women, in response to one hour of prolonged moderate intensity exercise. Twenty healthy women participated in this study. Ten were 25–35 (with a normal mean of 27.1 ± 1.3) weeks pregnant (P). The other 10 were nonpregnant (NP) and had normal menstrual cycles. Each subject was tested (a) for the determination of their maximal oxygen consumption (VO_2max), and (b) based on their VO_2max, and on a different day they performed 60 min of prolonged moderate intensity exercise, at 55% of their VO_2max, on a bicycle ergometer. On both occasions the subjects came to the laboratory 2 h postabsorptive and rested for 30 min. At that time a butterfly-type needle was inserted in an antecubital vein. Blood was collected before (0 min), during (at 15-min intervals), and after (15 min post) exercise and was analyzed for glucose and lactate. Additionally, electrocardiograms, fetal heart rates and maternal uterine contractions were monitored throughout the exercise testings. Glucose and lactate were determined by the glucose and lactate oxidase method respectively, by using a YSI model 23L Lactate/Glucose analyzer. This study and all the experimental procedures were reviewed and approved by the department of health services of the Los Angeles County Medical Center. The results of this study indicate that despite the similar intensities at which both P and NP women exercised, differences did occur in glucose homeostasis. Glucose levels in P women dropped to a lower (than NP) level, without establishing the steady state which was evidenced in the NP group, during the course of exercise (p < 0.05). The drop of blood glucose in the P group at the 45th and 60th min was not only significantly lower than the 0-min value but was also significantly lower than the 15-min value (p < 0.001). This indicates a two-level drop of blood glucose in the P group, which was not evidenced in the NP group. The lowest value of glucose in the P group was reached at the 60th min of exercise and was 3.66 ± 0.17 mmol/l (mean ± SE). (In other studies at similar levels of maternal blood glucose, which are produced by 24 h of fasting, the concentration of amniotic fluid glucose drops to half the amount which occurs at normal conditions.) Fifteen minutes after the end of exercise, the blood glucose levels of the P women remained low, while those of the NP had returned to pre-exercise levels. Lactate levels increased in both groups at the initiation of exercise, but dropped at a faster rate in pregnant women, reaching a significantly lower value by the end of exercise. The findings suggest that glucose falls and at a faster rate and to a lower level when compared to the nonpregnant controls throughout one hour of submaximal (55% VO_2max) cycling. Since the relative percentage of work load performed was similar between the P and NP group, it can be concluded that there appears to be an acceleration and/or exaggeration of the metabolic processes.

Supported by a University of Southern California Grant.

Total Body Protein Measurements by in vivo Neutron Activation Analysis: A Proposal for Monitoring Body Protein to Assess Nutritional Status and Body Fitness

Ion E. Stamatelatos[a], *Seiichi Yasumura*[b], *Dimitris Glaros*[a]

[a] Medical Physics Laboratory, Medical School, University of Ioannina, Greece;
[b] Medical Department, Brookhaven National Laboratory, Upton, N.Y., USA

Total body protein (TBP) is perhaps the most significant parameter characterizing the nutritional status of an individual. In critically ill patients it is important to measure TBP to monitor the effects of nutritional interventions, however, it is also useful to determine TBP in normal individuals to assess their nutritional status. Since the ratio of TBP to total body nitrogen (TBN) is fixed, TBP can be inferred by the measurement of TBN. Currently, prompt gamma neutron activation analysis (PGNAA) is the only established technique for measuring TBN in vivo.

We propose to establish such a PGNAA facility in Greece to measure TBN in vivo. The facility will include the use of two GBq (15 Ci) Americium-beryllium (^{241}Am-Be) radionuclide neutron sources placed one above and one below the patient scanning bed and housed within a graphite collimator and reflector assembly. Sodium iodide detectors will be positioned at ninety degress to the beam axis and bed. The subject will be scanned in five sections of 20 cm, 4 min per section, for a total scan time of 20 min.

Based on our previous work on PGNAA systems at the University of Birmingham, UK, and Brookhaven National Laboratory, USA, we estimate that the precision of the TBN measurement will be 2.5% for an effective dose equivalent of less than 0.15 mSv (QF = 10). This dose is sufficiently low that it will enable us to make TBN measurements in normal subjects.

The primary reason for having a PGNAA unit to measure TBN is to assess nutritional treatment protocols in seriously ill protein-depleted patients and thus the facility will be built within a clinical setting. This will provide the means to assess the nutritional status of patients and to evaluate the effects of treatment of malnourished individuals undergoing enteral and/or parenteral nutritional therapy. However, such a facility will also be available for the determination of TBP in the general population, with particular reference to athletes. This will add valuable information about the effectiveness of various nutrition regimens as well as providing an assessment of the effectiveness of training.

Dietary Manipulation Improved the Metabolic Response to Exercise

A. Szezesna, K. Wysocki
Department of Physiology, 'Jedrzej Sniadecki' Academy of Physical Education, Gdansk-Oliwa, Poland

This study investigated whether the combination of a starvation, carbohydrate-rich diet given after glycogen-depleting exercise and an alkalizing diet increased the exercise endurance in highly trained subjects. We expected that these nutritional strategies could be taken to cause: (1) an increase of glycogen stores in the ST and FT fiber of skeletal muscle and liver before the exercise; (2) a decrease of the rate of carbohydrate utilization during the exercise by increased use of fat as energy substrates, and (3) an increase of the buffering capacity in blood.

Materials and Methods. The subjects were six members (19–26 years of age and 64–126 kg weight) of the national championship judo team. Three preliminary testing sessions were performed. Two exercise tests on a cycle ergometer were performed prior to the main study. After one habituation experiment where the subjects were familiarized with experiment technique, the second exercise test prior to the procedure contained: 24 h starvation, intense twice-daily specific training, carbohydrate-rich diet and alkalizing diet (procedure S+C+A) and the main exercise test after procedure S+C+A were performed. The subjects performed an exercise test on cycle ergometer (Monark, Sweden) at a constant pedaling frequency at 60 rpm in which the initial work rate consisted of three minutes of pedaling on an unloaded cycle ergometer. Thereafter the work rate was increased to 3 W/kg. The subjects exercised 300 s. After 5 min cycling, the rate work was increased by 50 W every 1 min until exhaustion. Time to exhaustion and work output for the performance trials were calculated. During the test O_2 consumption (VO_2), respiratory exchange ratio (R) were recorded (using Spirolit, Germany), HR recorded from EKG. Additionally blood samples were collected before the exercise test and 2 min after ending the test. Blood pH, pCO_2, pO_2 were measured using blood gas analyzer ABL 3 (Radiometer, Copenhagen) and plasma HCO_3 and BE were calculated. During the exercise test after procedure S+C+A the men performed 14% more work (219.2 ± 74.4 vs. 187 ± 50.4 kJ), time to fatigue was significantly longer (683 ± 131 vs. 545 ± 80 s), R fell from 0.90 ± 0.14 to 0.82 ± 0.07, blood pH after exercise performed before and after procedure S+C+A was 7.14 ± 0.05 and 7.16 ± 0.05, respectively. In conclusion, the present results indicate that, as a result of applying the procedure, performance capacity and power output during training were enhanced in highly trained subjects.

Cost-Benefit Analysis of Investment in Athletics in the Framework of a National Development Plan

S. Theofanides
Panteion University, Athens, Greece

Investment in athletics (allocation of resources in terms of money, time, etc.) is evaluated ex post by the method of cost-benefit analysis.

A complete model must estimate the profitability of such investment from various aspects: (a) for the individual; (b) for his family; (c) for the enterprise or firm concerned; (d) for the society as a whole (national evaluation):

(I) The *stream of costs* of athletic investment include: the infrastructure, the cost of transport of athletes, the alternative cost (in the best alternative use of leisure time), additional expenses (for athletic shoes, etc.), the salary of the training personnel, the insurance cost.

(II) The *stream of benefits* of athletic investment include: the increase in the health of the body (fitness), the improvement in psychological behavior, the increase in the productivity of individuals, the positive image for the individual and society, the avoidance of undesirable activities (narcotics, etc.), generation of income (offer of athletic services), generation of employment opportunities (athletic press, television, training staff, gymnastic academies, etc.).

The quantification of the above streams of C and B for a period of analysis of 50 years shows that benefits far exceed the costs (in terms of discounting flows) and, therefore, investment in athletics is socially profitable. This favors investment in athletic activity in the National Development Plan.

Is the Type of Sports a Limiting Factor in Lipid and Lipoprotein Levels in Athletes?

Alexander Tsopanakis
Hellenic Sports Research Institute, Athens, Greece

A question is raised based on experimental results with respect to the ratio factor (RF) of lipoprotein concentrations, low density lipoproteins to high density lipoproteins (LDL/HDL) or total cholesterol to high-density lipoproteins (TC/HDL), which led us to propose that the RF is a useful parameter in assessing lipid adaptations to physical training according to different athletic specialties. The RF value is considered also to be a more useful parameter to assess the risk of CHD than the HDL value itself (Framingham study and others). Lately the above ratio factor seems to be correlated with certain stress factors and this is also under investigation by our group.

In this report we present the results and compare the lipid and lipoprotein profiles of elite athletes, divided in groups of 25 specialties forming 4 unities (n = 554) to a sedentary one (n = 63) as control. The sequence of athletic specialties is observed in order of increasing RF (TC/HDL) values. Control RF was used as limit ratio value (RF = 3.98). Endurance sports such as team games (football, basketball, volleyball) as well as short and long distance running and cycling show favorable high HDL and low RF values compared with sedentary people. In non-endurance anaerobic and strength sports such as wrestling, boxing and fencing, slalom, sprint and jumping, lipoprotein values are found to be nearer to controls (non-athletes). According to that classification, the first unity of athletic specialties (team games) includes tennis, basketball, volleyball, ping pong, handball, and water skiing. The second unity (aerobic disciplines) includes cycling, MD running (800–1,500 m), football, skiing, rowing, boxing, LD running ($>$ 3,000 m), while the third unity (anaerobic) includes judo, wrestling, SD running (400 m), jumping, throwing, swimming, sprint, water polo and fencing. Next to this unity lies the fourth (static endurance and power), i.e. sailing, weight lifting, car and motorcycle racing. The consequence of lipid ratio factor (RF) is further discussed with respect to lipid metabolism and exercise protection against atherogenesis.

Nutritional Evolution of Drug Addicts under Detoxification

P. Varela, A. Marcos, A. Casco, A. Requejo
Institute de Nutricion CSIC, Facultad de Farmacia, Universidad Complutense, Madrid, Spain

Since drug abuse has been pointed out to be associated with anorexia involving special food behavior and therefore to effect fitness, the aim of this study was to evaluate some aspects of nutritional status in 56 male heroin addict who voluntarily sought detoxification therapy, carried out over two different periods: (I) 1 month (n = 20), and (II) 4 months (n = 36), by determining the adjustment of the diet to RDA as well as anthropometric and biochemical parameters. The subjects were divided into two groups within each period, depending on the infection by human immunodeficiency virus (HIV): (1) HIV negative (I: n = 12; II: n = 22) and (2) HIV positive (I: n = 8; II: n = 14). Food intake, height, previous and actual weight, previous and actual BMI, weigh recovered, blood pressure, pulse rate, serum total proteins, albumin, cholesterol and triglyceride were measured. Energy supply for drug addicts in period I was over RDA, while these values were in the inferior borderline for the subjects in period II, besides in both periods energy supply was higher for HIV positive drug abusers. Moreover, calorie percentages supplied by proteins and lipids into the diet were also in excess in both periods. Likewise, these outcomes take place also in the Spanish population. A trend to increase lipid intake was observed in positive drug addicts, which is linked to a higher cholesterol intake of the subjects in period I. In spite of these results, weight recovery values for positive heroin abusers were lower than those found in the serum-negative group in period I, who reached similar values in both periods. Total serum protein values remained unmodified in all groups tested, but protein pattern suffered changes as a consequence of either HIV infection or the period of drug detoxification. Thus generally, albumin levels for drug addicts in period I were higher than in period II at the expense of a decrease in gamma-globulin, which was more important than in period I. These protein values led to an increase in albumin/globulin ratio, which was higher due to the HIV infection. A trend to increase cholesterol levels together with a decrease in triglycerides was shown in drug addicts in period II in comparison with those from period I, although no significant differences were found between them.

In view of these results, we might conclude that all heroin drug abusers tested in this study showed an acceptable nutritional status. However, the period of detoxification and infection with HIV modulate some of the nutritional aspects.

Impact of Supplementation on Performance (Sports) of Adolescent Girls – Rural, Urban

Kunjamma Varghese
Avinashilingam Deemed University, Coimbatore, India

This study was to evaluate and compare the impact of supplementation on performance (sports) of girls age 13–18 years in rural and urban areas, in and around Coimbatore District, Tamil Nadu, India.

Socioeconomic status and nutritional status including clinical assessment of 2,000 girls, 1,000 from urban and 1,000 from rural areas were recorded. They were grouped based on their age as 200 in each age group of 13 years, 14 years, 15 years, 16 years and 17–18 years, and out of the 5 groups in both rural and urban areas, 100 were selected (50 as control and 50 as experimental group) for both the urban and rural groups. The parameters selected to study the impact of supplementation of nutritious biscuits for a period of 2 months. Anthropometric parameters like height and weight, biochemical parameter haemoglobin, and performance testing parameters, such as agility, speed, strength (right arm, left arm), power, flexibility, muscular endurance, cardiovascular efficiency (physical fitness package components) were measured for all the 200 adolescent girls of urban and rural.

The following were the findings: The urban girls of the experimental group were taller and heavier, with mean height 148.5 cm and weight 43.7 kg. The lowest was the control group from the rural areas whose mean heights and weights were 145.76 cm and 32.5 kg, respectively. The haemoglobin levels were higher for the urban experimental group with 12.45 g/dl and lower for the rural control group, 11.63 g/dl. The performance of rural girls in parameters such as agility, speed, strength, power and flexibility were greater whereas muscular endurance and cardiovascular efficiency were higher for urban girls (53.5) which may be attributed to their higher haemoglobin levels. There was improvement in all the parameters studied due to supplementation of nutritious biscuits for 2 months, power improved in all the groups whereas the improvement is higher in experimental groups of urban and rural which is 0.06 and 0.04, respectively. Similar trend was observed in flexibility.

Though this study is only a limited one, it gives ample scope to speculate the possible role of nutrition in sports performance.

Body Mass and Mineral Losses by Sweat in Endurance Athletes

C. Wenk, W. Langhans
Laboratory of Human Nutrition, Federal Institute of Technology,
Zurich, Switzerland

In well-trained endurance athletes, one of the first limiting factors of performance is the water loss by sweat and by respiration. But also in everyday life important water losses have to be avoided. In contrast to the water losses by respiration the losses by sweat are followed by losses of minerals, mainly Na, K and at a lower level Mg, Ca and eventually Fe.

In 2 experiments with endurance runners over the distances of 30 and 51 kgm, we have tried to quantify the body mass and mineral losses of 17 (1st experiment) and 16 (2nd experiment) voluntary participants by the whole-body washing method. In both experiments two different beverages were tested. In the 1st experiment the content of the carbohydrates (CHO) was varied. Within a period of 1 month the participants had to complete in each experiment the distance twice in a cross-over design.

Description of the beverages: Experiment 1: treatment 1 = sports drink with carbohydrates and vitamins; treatment 2 = sports drink with carbohydrates, vitamins, and minerals (Na, K, Ca and Mg). Experiment 2: treatment 1 = beverage with 80 g CHO and minerals (Na, K, Ca and Mg); treatment 2 = beverage with 5 g CHO and minerals (Na, K, Ca and Mg).

The performance of the athletes was not influenced by the different beverages. Also the body mass losses and the serum concentrations of the minerals were not affected by the different treatments and the duration of the races. The homeostasis enabled constant metabolic conditions, even if no minerals were supplemented to the drinks. In the race over 30 km, the sweat losses amounted to 2.2 and 2.3% of the total body mass in treatments 1 and 2, respectively. Over the long distance of 51 km the corresponding values were 5.4 and 5.3%. Between the treatments no statistically significant differences of the mineral losses by sweating were detected. They amounted to approximately 490 and 650 mg/kg deltaBM for Na, 190 and 130 mg/kg deltaBM for K, 4.0 and 2.5 mg/kg deltaBM for Mg and 25 and 20 mg/kg deltaBM for Ca in the experiments 1 and 2.

The Beneficial Effects of a Kind of Chinese Tonic on Mice

Liu Yi, Li Liu Bai, Wang Hui Gi
Department of Nutrition and Food Hygiene, Beijing Medical University, Beijing, China

The Chinese tonic which has beneficial effects for health usually consists of purified Chinese traditional herbs. The purpose of this study was to evaluate the beneficial effects of a kind of Chinese tonic which consisted of *Lycium chinesis* and Gueihyuarn, two kinds of Chinese traditional herbs that are also foodstuffs. The results showed: (1) The use of this tonic helped the body weight of mice to rise, the mean gained body weight of experimental groups which were supplied with this tonic in different doses for 3 weeks were 8.8–9.5 g, while that of the control group was 4.8 g ($p < 0.05$). (2) The use of this tonic helped to increase the swimming endurance of mice in a flume with running water, the increase in value of swimming time of experimental groups was longer than that of the control group by 57.2–75.6% ($p < 0.05$). (3) The use of this tonic helped to increase the tolerance in absence-oxygen condition, the increase in value of survival time of the experimental groups was longer than that of the control group by 16.1–18.4% ($p < 0.05$). These results suggest that this tonic has a beneficial effect of promoting the health and sports ability of mice. This kind of Chinese tonic has the advantages of being both a herb and a food.

Subject Index